FROM BR TO BEECHING

VOLUME THREE

The Routes of t[illegible]l and Pepperc[illegible]

by PATRICK WHITEHOUSE and DAVID JENKINSON

CONTENTS

FRONT COVER

The pioneer Thompson Class A2/3, No. 60500 *Edward Thompson* itself, calls at St. Neots with a southbound stopping train on 31st May 1962. (Alan Tyson)

REAR COVER TOP

Peppercorn Class A2 No. 60528 *Tudor Minstrel* in charge of a freight train at Dundee on 28th August 1965. (Colour Rail)

REAR COVER BOTTOM

Peppercorn Class A1 No. 60156 *Great Central* leaves Grantham on 29th August 1961 with a southbound express from Newcastle for Kings Cross. (Colour Rail)

ISBN 0-906899-41-9

First Published 1990

PHOTOGRAPHS:
As credited.

Designed by
Barnabus Design & Print · Truro · Cornwall

Typeset by
TypeStyle · Truro · Cornwall

Printed by
Century Litho · Penryn · Cornwall

Bound by
Booths Bookbinders · Penryn · Cornwall

Published by
Atlantic Transport Publishers
Trevithick House · West End
Penryn · Cornwall TR10 8HE

Atlantic

AUTHORS' INTRODUCTION

In offering this, the third in our 'BR to Beeching' series, we move to one of the arguably less well publicised aspects of the early BR scene – the 90 post-Gresley 4-6-2s to LNER design. Again, we have obtained the help of acknowledged outside experts in the field, this time two of them. Peter Townend (one time shedmaster at Kings Cross) has provided a valuable appreciation of the Peppercorn engines (mostly based in England), while John Powell offers his usual pithy observations on the general situation north of Newcastle; we are deeply grateful for their efforts. It therefore only remains for us to thank the photographers past and present whose efforts have made this work possible and of course our usual picture researchers and data checkers, John Edgington and John Smart, for their continued stalwart work in helping sort through what was in this case an unusually rich collection of images.

We have also felt it helpful, before embarking on the major story, to devote a few pages to the evolution of the post-Gresley 4-6-2s, including a clutch of views chosen mainly to help our readers 'spot the differences'.

Patrick Whitehouse **David Jenkinson**

CHAPTER ONE

A NEW STYLE OF 4-6-2

a) Establishing the Type

The post-Gresley pacifics first appeared in the 6ft 2in driving wheel form, later known as the A2, and stemmed from Edward Thompson's wartime (1943) rebuilding of the pioneer Gresley Class P2 2-8-2s in an attempt to eliminate some of their problems – not least the long coupled wheelbase in relation to the twisting Edinburgh-Aberdeen route. A somewhat ungainly result emerged from Thompson's wish for both divided drive and equal length connecting rods for all three cylinders and this resulted in the outside cylinders being set further back – between bogie and leading coupled wheels – with unfortunate aesthetic and at times mechanical effects. These rebuilds were followed in 1944 with four similar examples described as 'modified V2s', having originally been ordered as Gresley 2-6-2s. At first, these ran with their ordered six-wheel tenders (soon changed for conventional 8-wheel 'pacific' tenders); but the main visual distinction of these engines from the P2 rebuilds was the early presence of large smoke deflectors in place of the small chimney 'shields'. This latter change was to be repeated in the main production batch of Thompson A2s, built new as such from 1946. These also differed in having flat-fronted cabs.

The versatility of a 6ft 2in pacific was obvious to A. H. Peppercorn when he took over and he 'cleaned up' Thompson's design to good effect with his first A2 at the end of 1947. The divided drive was maintained but equal length connecting rods were abandoned and as a result, the outside cylinders went back to their 'normal' position – with beneficial visible and practical effect. This also meant that the smoke deflector shields on the Peppercorn A2s had a larger sized rear 'cut-away' to facilitate access to the footplate-top above the cylinders. Angled front cab windows were also reinstated (though not the full 'V' front). In the final LNER classification, the Peppercorn engines were regarded as the 'standard' type, the earlier Thompson engines being given the familiar sub-class part numbering (see below).

More numerous than the 6ft 2in engines were the 6ft 8in A1 series, whose origins went back to 1945 with the highly controversial Thompson rebuild of Gresley's pioneer 4-6-2 No. 4470 *Great Northern*. We shall not be drawn into the emotional argument regarding his choice of engine – for there is some evidence that Thompson was genuinely seeking to effect improvements and simplification of the more complicated Gresley derived motion in the difficult war years – but like the rebuilt P2s, the visual result was poor; some even called it a disaster and this deterioration in looks often made it even less acceptable given its origins. The ungainly gap between bogie and coupled wheels appeared again (for the same reason), this time linked with an unhappy stovepipe chimney and an oddly short design of cabside sheet, to produce a less than harmonious ensemble compared with the original engine.

Happily, no more Thompson rebuilds of Gresley's 6ft 8in pacifics appeared and the larger wheeled express type was not repeated until Peppercorn introduced his own A1 design in 1948. As with the A2s, he modified Thompson's concept into a fine and balanced looking engine of no mean performance potential. All told, 49 of these new engines were built (all by BR, of course) and as with the A2s, the Peppercorn design was regarded as the 'standard' A1 and the solitary *Great Northern* became a sub-part. In spite of their obvious differences from their Gresley predecessors, the Thompson and Peppercorn engines were both very much in the LNER design tradition and in the shape of the main batch of A1s, had the distinction of being the last new type of express steam engines to be built in Britain to a wholly company-inspired design.

We give a breakdown of types below (using BR numbers) and further detail differences between the various classes are provided in our first few views.

Summary

Thompson and Peppercorn Class A2 4-6-2s

BR No. Series	Designer	Date Introduced	LNER Class/Remarks
60500	Thompson	May 1946	A2/3, built new
60501-6	Thompson	January 1943	A2/2, rebuilt ex-P2
60507-10	Thompson	May 1944	A2/1, 'modified' V2
60511-24	Thompson	May 1946	A2/3, built new
60525-39*	Peppercorn	December 1947	A2, built new

* Nos. 60526-39 built by BR, Nos. 60527-31 having 'E' prefix to LNER number when new.

Thompson and Peppercorn Class A1 4-6-2s

BR No. Series	Designer	Date Introduced	LNER Class/Remarks
60113	Thompson	September 1945	A1/1, rebuilt ex-Gresley
60114-62	Peppercorn	August 1948	A1, built new

In order to cover the story of the Thompson 4-6-2s properly, it is first necessary to illustrate their predecessors, the class P2 2-8-2s from which they were rebuilt. No. 2001 *Cock o' the North* was built in May 1934 as shown here with rotary cam poppet valve gear. No. 2002 *Earl Marischal* (10/34) was similar but with Walschaerts/Gresley valve gear while the remaining four locomotives were built in 1936 with streamlined casing at the front end similar to the A4s (eg No. 2004 *Mous Meg* illustrated). All the P2s except No. 2005 had double chimneys. No. 2002 was fitted with extra smoke deflectors in April 1935 but was later streamlined (as per Nos. 2003-6) in October 1936. No. 2001 received Walschaerts valve gear and 'A4' type streamlining in April 1938. The P2s were, perhaps, the ultimate expression of Gresley's 'horses for courses' philosophy, a capable if flawed design for working the Edinburgh-Aberdeen line. A numerically small class, with careful handling and maintenance required, they had little place in a wartime environment and rebuilding to 4-6-2 form took place between January 1943 and December 1944. (Ransome Wallis Collection/NRM)

A2/2 No. 60502 *Earl Marischal* (rebuilt 6/44-7/61) gives a fine impression of the rebuilt P2s as it stands outside the old Lancashire & Yorkshire shed at York, so called because of the L&YR locomotives that visited it for servicing – it closed in 1933 but remained in use as sidings and a turning point until the mid 1950s. By this time *Earl Marischal* had been fitted with a lipped chimney, improving its appearance somewhat; but as with all the other A2/2s it retained the diminutive smoke deflectors fitted after rebuilding. No. 60502 was a York based engine and is probably standing pilot awaiting a call to replace an ailing locomotive on an up train. Note the inspection saloon in the shed, it is No. 21661 a NER rebuild of a Stockton & Darlington vehicle now preserved privately on the Keighley & Worth Valley Railway. (Eric Treacy/Millbrook House Collection)

Above: As part of his early experimentation, Thompson caused four, already ordered Gresley V2 type 2-6-2s to be built as 4-6-2s, embodying his new ideas. On appearance, they much resembled the rebuilt P2s, save for the very obvious newly styled smoke deflectors fitted soon after building. They were the first LNER engines to receive this form of embellishment as standard equipment (which they retained to the end) and this c. 1956-7 view shows one of them, No. 60508 *Duke of Rothesay* heading north from York. The picture was taken right at the transition point between old/new BR tender emblem (the old version is seen here) and between the original (red/cream) and later (maroon) carriage liveries. (Eric Treacy/Millbrook House Collection)

Left: The 'production' series of Thompson A2s, built as such, were very similar to the 'modified' V2s and came into service with 'round' domes and stovepipe chimneys. This initial configuration is represented here by New England (35A) based A2/3 No. 60523 *Sun Castle* (8/47-6/63) as it passes Holloway South (the down side box is just out of shot to the right) in 1951-52 with the 10.05 am relief 'Flying Scotsman'. The leading coaches are for Aberdeen and consist of a pair of Thompson pressure ventilated vehicles, still in 'teak' livery and originally built for the first post-war LNER 'Flying Scotsman' services. Brake third No. E1904E is followed by a shorter-underframed composite. These coaches were an interesting Thompson innovation in design terms, having a series of intermediate cross-vestibules as well as conventional end-entrances, thus ensuring that no passengers were more than one or two compartment lengths away from an outside door. This made it difficult to get the normal seven compartments in the composites, hence their shorter 58ft length and a reduction to six compartment layout. Behind these coaches are the newly-introduced BR Mk. 1s. The 10.05 was a regular turn for the New England A2s. Class A2/3s were allocated there from December 1948 and a later view of the same train can be seen on page 15. (Eric Treacy/Millbrook House Collection)

Above: The first A2 6ft 2in pacific No. 525 *A. H. Peppercorn* (12/47-3/63) stands at Kings Cross on 22 March 1948. It is awaiting the 6.05 pm return working to its home depot Doncaster. No. 525 was the only Peppercorn Pacific to appear before Nationalisation (though No. 526 also came into service without BR number marking) and is seen here in the final version of LNER apple green livery, although the nameplate has a red background rather than the usual black. The locomotive spent most of its career north of the border working from Aberdeen (Ferryhill) depot, following transfer from New England shed in August 1949. Note the much tidier and more compact arrangement of the 'front-end' and the revised smoke deflector shape compared with the earlier Thompson examples. (C. C. B. Herbert Collection/NRM)

Right: The pioneer GNR class A1 4-6-2 No. 4470 *Great Northern* built in 1922 was rebuilt by Edward Thompson in September 1945. It is seen here in the locomotive yard at Kings Cross station around October 1945 with high cab, stovepipe double chimney and no smoke deflectors. The livery is royal blue with red lining; note the shortened NE on the tender a wartime economy. In this form, regardless of the rights and wrongs of the rebuilding, it is not difficult to see why the appearance of the reconstructed *Great Northern* was compared unfavourably with the graceful lines of the original Gresley design. (C. C. B. Herbert Collection/NRM)

b) The Peppercorn A1s and A2s – an appreciation

*by **Peter N. Townend***

If any design of express passenger locomotives deserved the classification A1, the post-war LNER Peppercorn pacifics did. Together with the smaller 6ft. 2in. driving wheel but basically similar locomotives these were the last of the LNER pacific designs which brought together all the detailed refinements developed since the original Gresley pacific emerged from Doncaster Works at the end of March 1922. Gresley had anticipated, after the success of the streamlined trains before the war, that heavier loads at higher speeds would require a more powerful pacific and had rebuilt the experimental engine No. 10000 with 19in cylinders and a 250psi boiler fitted with a 50 sq ft grate, somewhat larger than the earlier 41.25 sq ft of the 4-6-2s. The unusual 4-6-2-2 wheel arrangement was retained in this rebuild, classified W1.

The large fleet of Gresley three cylinder pacifics and V2s had coped extremely well with the very heavy wartime trains without resorting to double heading but in 1945, when the orders were placed for the new A1 and A2 class locomotives, the basic dimensions were of the more powerful rebuilt W1 No. 10000. A total of 49 streamlined A1 4-6-2s were eventually ordered, to be capable of working 500 ton trains at 60 mph and 600 ton trains at 50 mph average speeds over the East Coast main lines. The tractive effort of the A1 class equated to 37,397 lbs. at the usual 85% of the boiler pressure and the A2 was 40,430 lbs. making it nominally the most powerful express engine in the country. The 15 A2s were originally part of an order for the earlier Thompson A2/3s, but with the retirement of the CME, it was possible to revert to the more conventional cylinder arrangement and the 15 A2 class engines were constructed in 1947-48.

The building of the A1 class immediately followed, 26 being constructed at Doncaster and 23 at Darlington in 1948-49. All were fitted with rocking grates and hopper ashpans, also electric lighting which was not always maintained in working order. The crankpin sizes were larger than the sizes used on the earlier Gresley pacifics and the bogie design was changed from the centre bearing type to one with side bearers. The external appearance of the A1 was changed by a decision of the Railway Executive not to permit streamlining which had been authorised by the LNER at an additional cost of £500 per locomotive. The LNER had however found the streamlined front to be the most effective form of smoke deflection after some years of trying different methods and fitting of double Kylchap blastpipes to all the A1 class made some form of smoke deflection essential, so deflectors were fitted from the footplating to both classes. Only the last of the A2 class was fitted new with a double Kylchap arrangement, although five more were provided when in 1949, MLS multiple regulators and steam dryers were fitted experimentally. The remainder of the class retained single chimneys and resulted in complaints about their steaming while this did not compare with the Kylchap arrangement.

The tenders fitted to both classes incorporated water sprays and steam brakes instead of vacuum which resulted in the A1 class not being able to be fitted with corridor tenders by straight exchange; in consequence they could not work the non-stop trains between Kings Cross and Edinburgh. Both classes were thoroughly tested in trials using the dynamometer car and the results were very satisfactory. There was very little difference between the two classes of locomotive, the A1 working to Leeds and back with around 500 tons produced a coal consumption of 40.2 lbs/mile – 1.8% less compared to the A2 in terms of drawbar horsepower hours. Working 600 ton trains between Kings Cross and Grantham and back, the results were reversed with the A1 burning 47.5 lbs/mile – 0.6% more than the A2 per drawbar horsepower hours.

Compared with the double Kylchap A4 used on 500 ton trains between Kings Cross and Leeds during the 1948 Interchange Trials, the A1 burned 1.3% more coal and the A2 1.8% more coal in terms of drawbar horsepower hours. The A4 had produced the best results of any locomotive tested in this respect, but the A1 and A2 could be expected to burn more coal with the larger grate anyway. On one particular journey during the tests between Leeds and Kings Cross with 490 tons, the A1 burned only 33.86 lbs. coal/mile (or 2.694 lb dhp. hour) and it is doubtful if any other pacific equalled this result under test in this country. Some higher power outputs were produced than during any of the interchange trials and the A1 produced 2108 dhp at 40% cut off and 61.2 mph, whereas the A2 attained 2138 dph at 45% cut off and 63.2 mph.

The A1s quickly took over much of the heavier mainline work and after through engine workings were established between London, Newcastle and Leeds, soon after their introduction, the allocation of the A1s was settled at 12 at Kings Cross, 3 at Doncaster, 3 at Grantham, 6 at Copley Hill, 14 at Gateshead, 3 at Heaton, 3 at York and 5 at Haymarket. Through engine workings were withdrawn (in favour of grouping classes together) in 1951 so that men could have their regular engines. Kings Cross lost its A1s mainly to Grantham in exchange for A4s. Copley Hill engines also had their regular crews on A1 locomotives, but Gateshead tended to allocate selected engines to specific workings, and two of the last five of the class fitted with roller bearings, allocated to Gateshead, frequently worked overnight on a lodging turn to Kings Cross returning with the 22.15 'Night Scotsman' the following night. The two engines ran on this two- day cycle with remarkable consistency and reliability, working the heaviest train of the day out of Kings Cross.

Later in the decade, the Eastern Region A1s were concentrated at Kings Cross and 12 A1s for a time worked alongside 19 A4s and 12 A3s. Many of the A1 and A4 engines were regularly manned by two crews and used on increased lodging turns to Newcastle. By then the policy of increasing speeds with lighter trains was well under way on the East Coast mainlines and it is doubtful if the larger grate of the A1 class was fully justified in these circumstances. The locomotives spent much of their working lives working well within their capacity.

The A2 class, of course, eventually became allocated to depots serving the Edinburgh and Aberdeen line, only four of the 15 built working outside Scotland, and only one of these working from New England in the Eastern Region. Both classes ran higher than average mileages between classified repairs and in everyday service required little attention with a lower than average failure rate. During 1953-55 the selected A1 class engines averaged a total of 8.53 pence per mile compared to 12.70 for an LMS 'Duchess' for example. Such a figure alone warranted their A1 classification. Neither class A1 or A2 were called upon to break any speed records but 100 mph was attained occasionally in service. Perhaps drivers were not keen at going much over the 90 mph track limit on locomotives that did not ride quite as steadily, as their predecessors, but nevertheless net times less than the pre-war 'Coronation' were achieved by a Kings Cross driver from York to Kings Cross with the A1 standing pilot when the diesel failed at York; the whole of the 26 minutes lost were more than recovered.

In 1955, a Grantham driver Royce showed what his A1 could really do by starting his engine from Grantham to such purpose (during a brake test run with a 462 ton train) that the engine averaged an equivalent drawbar horsepower of 2295 between milepost 100 and 99, attaining 90 mph by Corby. This was probably the highest horsepower produced by any of the Pacifics. Looking through the logs of runs recorded by Michael Joyce on his regular journeys between Kings Cross and Leeds (or vice versa), the remarkable feature was that they were all excellent runs with the A1 class engines. It is for this consistency of performance, daily high mileages and at a very low cost in maintenance that the A1s and A2s deserve to be remembered. And they were not only seen on former LNER metals either as the views on page 58 reveal.

DNT

Right: Haymarket based Peppercorn A1 No. 60152 *Holyrood* (7/49-6/65) leaves Edinburgh Waverley with a midday express for Carlisle via the Waverley route in the early 1960s. Note the lower position of the numberplate on the smokebox door hinge allowing the top lamp iron to be fitted further down. All five Haymarket A1s were transferred to St. Margarets in September 1963 and only No. 60152 survived beyond the end of the year. It moved to York in September 1964. (Eric Treacy/Millbrook House Collection)

CHAPTER TWO

KINGS CROSS TOWARDS NEWCASTLE

Like most 4-6-2s of LNER design or origin, the Thompson and Peppercorn pacifics were very much creatures of the demands of the East Coast Main Line. This book must therefore reflect this basic fact; but the physical nature of the ECML was not consistent. South of Doncaster it was predominantly easy, near-straight in alignment and in terms of gradients, had little of note to cause real trouble save for the celebrated climbs out of Kings Cross to the so-called 'Northern Heights' of the London basin – generally regarded as terminating somewhere near Potters Bar – and further out (in both directions) to Stoke Summit, c.100 miles from London. Even the latter was approached by the sort of gradients (1:178; 1:200) which the GWR west of Exeter would regard as no more than a 'hiccup'!

The route continued much the same almost as far as Newcastle, with the York-Darlington 'racing' stretch being perhaps the most celebrated of the high speed sections. In consequence, the bulk of railway photographers tended to concentrate efforts on the rather modest stretches of the line where engines might be expected to be working harder.

North of Newcastle, the route underwent a quantum change. The gradients became more severe and, especially north of Edinburgh, so to did the curvature of the main line itself. In this context, some interesting parallels can be drawn with the West Coast Main Line north of Lancashire – a subject to which we hope to return in this series – but in the purely East Coast context, it had its own effect on the nature of the motive power needed. Indeed, the Gresley P2s (the progenitors of the first Thompson pacifics) were themselves designed to cope with this very fact.

It will therefore occasion no surprise that we have elected to divide our main coverage of the work of these engines into the two parts of the main route which had their boundary somewhere near Newcastle. We shall, of course, divert from the main line from time to time, but we start our story, quite properly and predictably, at Kings Cross – not, perhaps, the most outstanding of London's termini, but without doubt one of its most interesting.

Right: Night scene at Kings Cross: Class A1 No. 60124 *Kenilworth* (3/49-3/66) of Gateshead shed (52A) awaits the right away at Platform 8. Just to the right of the smoke deflector above the signals are a pair of two way roller blind indicators, one showing Platform 10 for an incoming engine or empty stock train. (George Heiron)

a) Around Kings Cross

Right: Rail exits in the Kings Cross area.

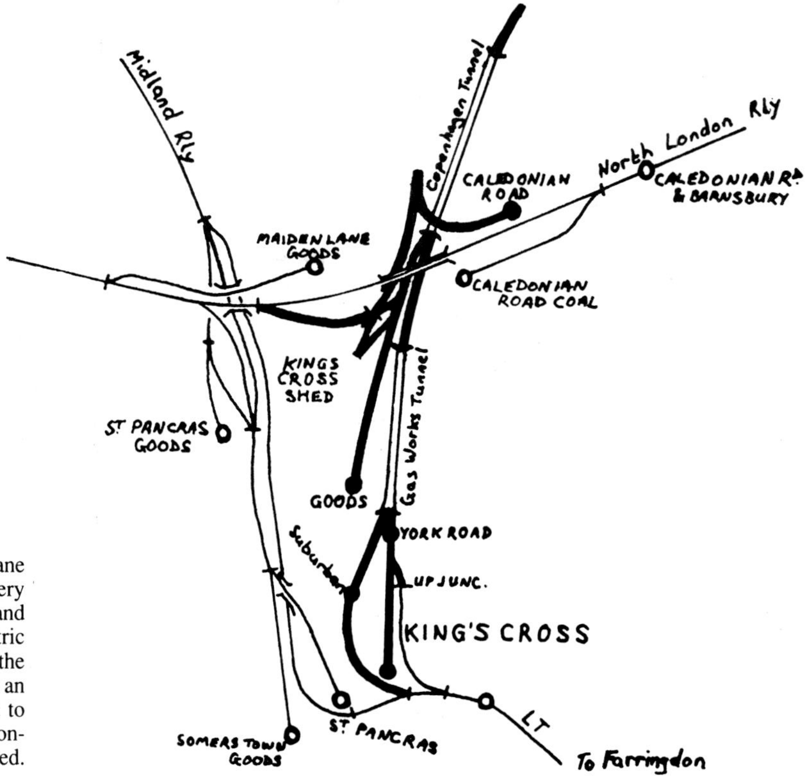

Below: A1 No. 60158 *Aberdonian* (11/49-12/64) takes water from a crane at Kings Cross in the early 1950s – the engine is now in BR green livery with the first tender emblem. The tender carried 5000 gallons of water and 9 tons of coal. Note the snaphead rivets used by Doncaster and the electric light system on the rear of the tender, similar in layout to that fitted to the diesel electric shunters. Note too the large solid fuel stove (fitted over at an alarming angle at the base of the water column). These were designed to prevent the columns freezing in cold weather; many railways used a conventional brazier but here, a closed-in arrangement was preferred. (Eric Treacy/Millbrook House Collection)

Above: A1 No. 60156 *Great Central* (10/49-5/65), viewed from above whilst being turned on the 70ft vacuum operated turntable at Kings Cross station. The double chimney, superheater header covers and anti-vacuum valve are all clearly shown, as is the hand painted Great Central Railway coat of arms on the nameplate – a feature of all the A1s which carried former railway company names. The smoke deflectors on the A1s were undoubtedly an asset both operationally and visually, in the latter case countering the fact that the double chimney was set well forward on the long smokebox. This view also clearly shows the practical reason for the re-styled Peppercorn smoke deflector compared with its Thompson predecessor. The more forward positioning of the cylinders on the Peppercorn engines (both A1 and A2) meant that the lower rear section of the deflector plate needed to be cut away more extensively to allow ready access to the outside steam pipe and footplate-top lubricators. (Eric Treacy/Millbrook House Collection)

Left: After only two months in traffic, see page 7, Thompson's rebuild of No. 4470 *Great Northern* entered Doncaster works in order to have altered some of the more unsatisfactory details. Large smoke deflectors were added and the cab was rebuilt on more conventional lines. This view taken at Kings Cross around 1948 shows these modifications. By this date it has been renumbered 113 under the Thompson post-war scheme and repainted in LNER post war apple green livery. (C. C. B. Herbert Collection/NRM)

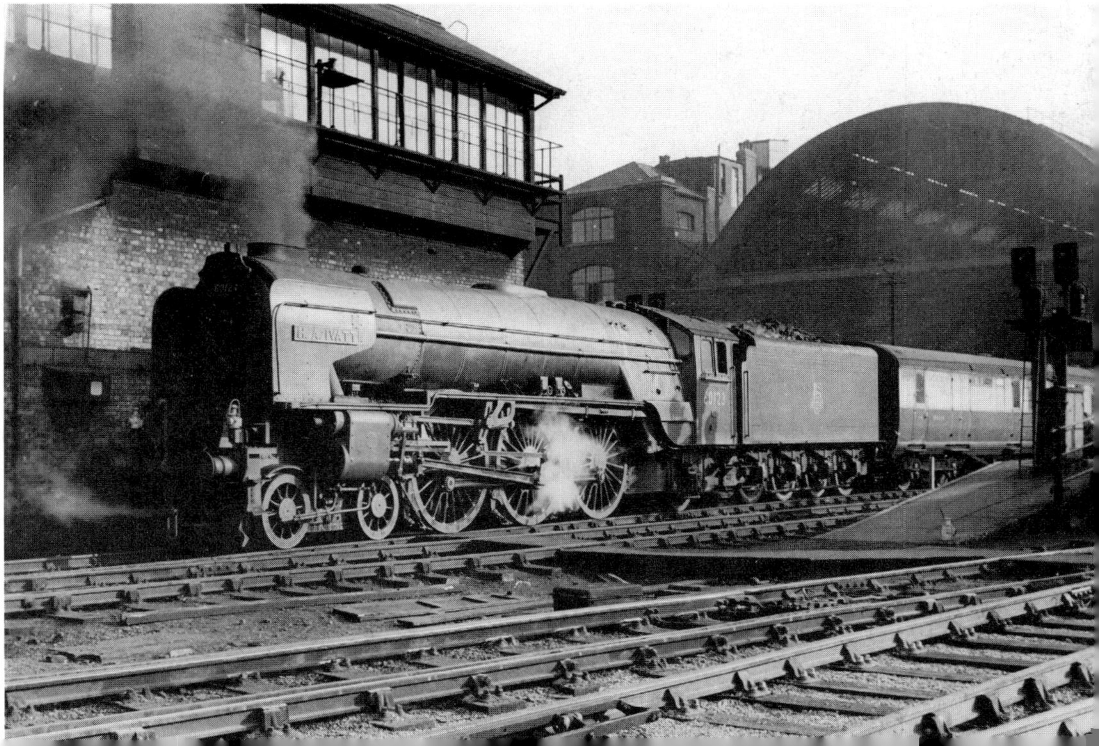

Right: In this early view (note the stovepipe chimney), blue-liveried class A1 No. 60123 *H. A. Ivatt* (2/49-10/62) stands at Kings Cross Platform 8 with the 3.45 'West Riding' for Leeds and Bradford in 1050 or 1951 – the locomotive was named in July 1950. The 'West Riding' was introduced in 1949 and included in its make-up, three pairs of articulated twins from the pre war 'West Riding Limited' streamliner, now painted in crimson and cream but retaining the aluminium trim from their two-tone blue days. This was a regular return duty for the Copley Hill (37B) A1s after working an early morning express from Leeds; in this case, the Copley Hill crew would have brought the up 'Yorkshire Pullman' into Kings Cross earlier in the afternoon, though the crew and locomotive worked different down trains on their way 'home'. (C. C. B. Herbert Collection/ NRM)

Right: Last years of steam at the Cross. A very shabby looking Thompson A2/3 No. 60513 *Dante* (8/46-4/63) of New England shed leaves Platform 2 at Kings Cross c.1962 with an evening local train, probably for Peterborough. Note the very complicated track layout, including many slip points. Another interesting (and long gone) feature is the suburban platform on the extreme left. This was used by trains bound direct for the City, which gained access to and from the Metropolitan widened lines via a very noxious tunnel, thus necessitating condenser-fitted locomotives. In the opposite direction, they re-emerged at Kings Cross by an equally subterranean route which entered the station on the extreme right, outside the main train shed and not quite visible in this view – but see map p.11. (Eric Treacy/Millbrook House Collection)

Left: The last Thompson Pacific, A2/3 No. 524 *Herringbone,* was completed at Doncaster in September 1947 and, having spent the majority of its life at York, was withdrawn from Polmadie shed in February 1965. It is seen here at Belle Isle climbing the 1:107 out of Kings Cross with a down express (possibly 'The Norseman' a summer boat-train for Newcastle Tyne Commission Quay connection with ferry services for Norway) circa 1957. The locomotive is fitted with an A1 pattern boiler (diagram 118) – note the 'banjo' steam collector and also the lipped double chimney – c.f. page 6. The coach livery is a mixture of the old red and cream (1948-56) and the new maroon (1956-65). St. Pancras station tower can be seen above Gas Works tunnel. To the right of the main lines is Kings Cross Goods depot and the roads leading to 'Top Shed', the celebrated servicing point for East Coast express locomotives. (Eric Treacy/Millbrook House Collection)

Right: A1 No. 60139 *Sea Eagle* (12/48-6/64) appears to be struggling as it climbs the 1:107 at Belle Isle with a morning express for the north despite the fine weather c.1955. The locomotive is in almost exactly the same position as No. 60524 in the previous view – note the two ground signals and catch point alongside the engine and tender. This once active location is now largely derelict, save for the main running lines to the terminus of course. The exit from Kings Cross goods was controlled by the Goods & Mineral Junction signalbox, seen above the tender, while the viaduct behind carried the North London line from Broad Street to Chalk Farm. No. 60139 was based at Copley Hill (37B) at this time so the train may well have been a return working to the West Riding. (Eric Treacy/Millbrook House Collection)

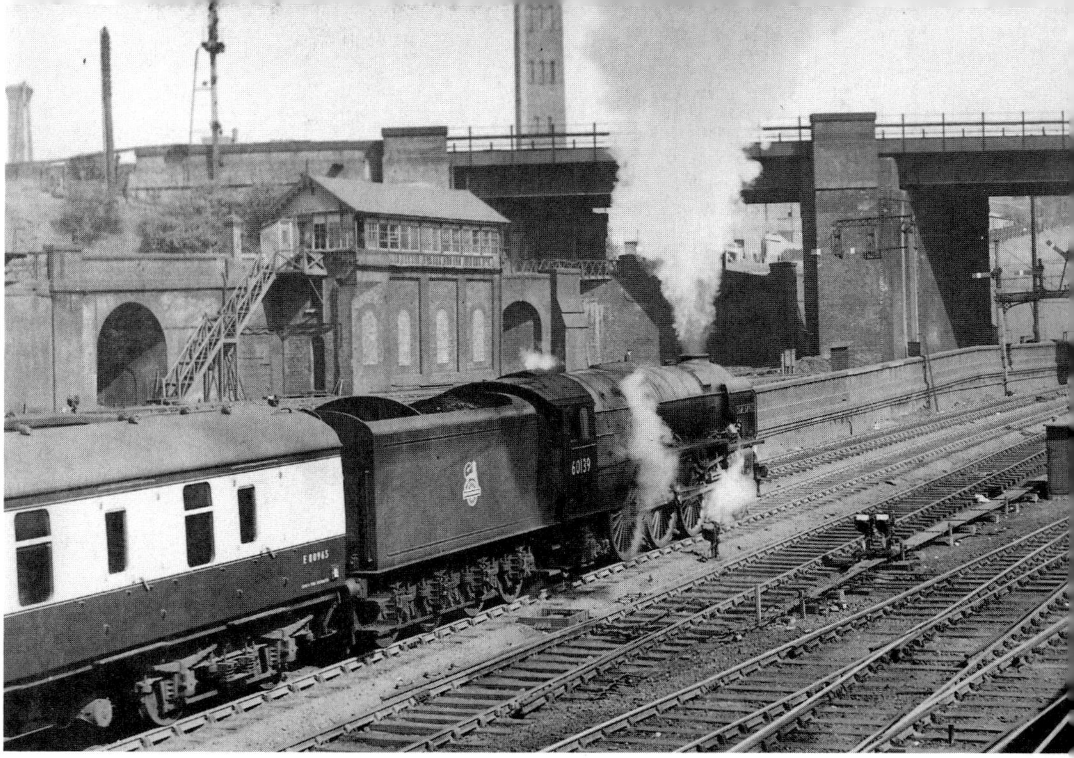

Right: A few yards further up the bank and beneath the 'North London' viaduct, Thompson Class A2/3 No. 60513 *Dante* (8/46-4/63) approaches Copenhagen tunnel with the down relief 'Scotsman' c.1954. This New England (35A) allocated engine will work the train to Peterborough. The first two coaches, Thompson brake third No. E1903E and BR Mk 1 composite, form the Aberdeen through portion; a slightly earlier version of the same train in charge of the same class of engine can be seen on page 6 but it should be noted that by comparison with the p.6 view, the Thompson engine seen here has received a boiler with a 'banjo' steam collector and lipped double chimney.

To the right, 'condensing' N2 0-6-2 tank No. 69579 (2/29-9/62) waits for a path down to Kings Cross, the inverted destination board on the smokebox inferring that it will soon be in charge of a set of quad-arts (four unit articulated suburban coaches) bound for Hertford. Beyond the viaduct are the goods lines into Kings Cross, which have a separate tunnel bore whilst above the brick retaining wall is the steeply graded branch to Caledonian Road (GNR) goods station. (Eric Treacy/Millbrook House Collection)

b) Langley Troughs to Grantham

Above: The up morning 'Talisman' on Langley troughs in 1957 headed by A1 No. 60125 *Scottish Union* (4/49-7/64). There were six sets of watertroughs on the East Coast main line, four on the GNR section at Langley, Werrington, Muskham and Scrooby and two on the former NER lines at Wiske Moor and Lucker. The NBR section did not have any troughs.

The morning 'Talismans' were introduced in the summer of 1957 to complement the 4 pm departures from Kings Cross and Edinburgh Waverley of the same name inaugurated in the autumn of 1956. These trains were relatively lightweight formations by customary East Coast standards but this permitted faster timings as a sort of half-way house towards the greater accelerations of the succeeding 'Deltic' and HST eras. Interestingly, the London Midland did much the same sort of thing with its 'Royal Scot' and 'Caledonian' services to Glasgow at much the same time. The 'Talisman' service was normally the preserve of Gresley A4s, so it is possible that Grantham was called upon to provide a replacement locomotive for the last leg of the journey. The eight coach set is a mixture of BR Mk I and LNER stock, all but one in 1956 crimson lake livery. The third and fourth vehicles are an articulated twin open first set from the pre-war 'Coronation' streamliner; behind these are two Thompson catering designs. (Ransome Wallis Collection/NRM)

Right: A2/2 No. 60506 *Wolf of Badenoch* (rebuilt ex-P2 5/44-4/61) stands at its home depot of New England, Peterborough (35A) on 22nd May 1955. It is carrying an A1 diagram 118 boiler but still retains the original 'stovepipe' chimney – c.f. page 5. The 50 sq. ft. grate area of the P2 boiler was retained by the rebuilds and used on all subsequent LNER Pacifics except the A2/1s which, of course, used the modified V2 boiler – see page 6.

The Scottish-inspired name clearly recalls its P2 origin in 1936 as one of Gresley's six 2-8-2s – an attempt to find a solution to the difficult Edinburgh-Aberdeen route – see also page 3. When rebuilt as A2/2s, they did not long remain in Scotland. It is interesting to note that this first series of Thompson A2s did not receive full size smoke deflectors and it is also worth remarking that on reclassification, though preceding the modified V2s in time terms, the P2 rebuilds were actually designated Class A2/2, rather than being given the more logical A2/1 classification which was allotted to the modified V2s. Behind the locomotive is the old coaling stage showing the old two-wheel coal tubs – the top of the replacement concrete coaling plant can be seen above. This depot housed around 150 steam locomotives at the time, including 9 Pacifics, all Thompson 6ft. 2in. types. (John Robertson Collection/ Colour Rail)

Left: The last LNER-designed Pacific to be built, No. 60162 *Saint Johnstoun* (12/49-10/63) of Haymarket shed (64B), rattles down the main line near Great Ponton with a partially fitted freight in the early to mid-1950s. The train is on the downhill stretch between Stoke Tunnel and Grantham and the engine was probably running in, following an overhaul at Doncaster works. Note the gated siding to the right owned by the Barnstone Cement Co. and the new platelayers' hut on the left. (Eric Treacy/Millbrook House Collection)

Left: Blue A1 No. 60148 *Aboyeur* (5/49-6/65) passes Grantham south box as it starts the climb to Stoke with the up 'Northumbrian', probably in the spring of 1952. This was a Newcastle-Kings Cross train which was named by BR in 1949 and already it contains a large proportion of BR standard coaches, introduced from 1951. The engine was one of ten allocated to Grantham after September 1951 when there was a major reorganisation of East Coast main line locomotive working, allowing most diagrams to be 'out and home' turns for both engine and crew.

It was at about this time that the A1s found themselves becoming the mainstay of most of the principal East Coast express steam workings. Many of the earlier Gresley 4-6-2s (save for those A4s reserved for the non-stop) were not always in the best of order and it was not until the later 1950s that they began to come into their own again (usually with double chimneys and other changes). Meantime, the East Coast had good cause to bless Peppercorn's new engines. One of the authors was a regular user of the route at this time (1952-6) and can scarcely recall an important train which did not have an A1 at its head. (Eric Treacy/Millbrook House Collection)

Below: A1 No. 60124 *Kenilworth* (3/49-3/66) allocated to Gateshead depot (52A) awaits its next call to duty at Grantham shed in the early 1950s. The position of the shadows suggests a mid-afternoon timing for this view and it is a fair supposition that *Kenilworth* had worked a train south from Newcastle earlier in the day and was now waiting for the balancing working of an 'out and home' turn. A consequence of this form of operation was that Grantham would regularly play host to a large number of distinguished visitors in addition to its own allocation, one of which, Gresley 02 2-8-0 No. 63923 (5/21-12/62), stands alongside. In January 1954 the shed had an allocation of fourteen 02s, mainly for working iron ore trains from High Dyke sidings to Scunthorpe. (Eric Treacy/Millbrook House Collection)

These two views show the somewhat infamous Thompson A1/1 No. 60113 *Great Northern* leaving its home town of Grantham (35B), where it was allocated from September 1951 to September 1957. It shared duties with the depot's Peppercorn A1s but proved unreliable and latterly was used mainly as Grantham pilot.

In the first picture (right), *Great Northern* heads a heavy up express out of the station c.1953. It is in BR green livery and now has a lipped chimney. Note the tall home signal to allow sighting over the footbridge and repeated below. Lofty signals were a real Great Northern Railway tradition although by now upper quadrant arms had replaced the original somersault type. Unusually there are two full brakes in the train.

Left: In the second view, No. 60113 leaves the station in the opposite direction with an afternoon down express in the mid 1950s. Note the new nameplate, fitted in May 1951 and incorporating the Great Northern Railway coat of arms. (Eric Treacy/Millbrook House Collection – both)

c) Doncaster, Leeds and the West Riding

Above: Grantham (35B) based Peppercorn A2 No. 60533 *Happy Knight* (4/48-6/63) pauses at Doncaster with the southbound 'White Rose' in 1955 or 1956. This was a Bradford, Leeds and Kings Cross service leaving Leeds at 3.15 pm, named by BR in 1949. The train is a mixture of BR Mk I and LNER-designed Thompson stock – a common combination in the early/mid 1950s. The exact formation is obscured by the platform canopies but the second vehicle is one of the relatively uncommon (and very spartan!) Thompson open thirds, later seconds. Note the canopies over the up and down main platforms and the concrete lamp posts, the nearest has a station nameboard attached. Much of this has now vanished (1990) but the basic configuration of the station at Doncaster remains much as shown here. (Eric Treacy/Millbrook House Collection)

Below: North of Doncaster, the main East Coast line followed its familiar fairly level and straightforward alignment, but those trains which 'turned left for Leeds' had a rather more demanding route for their last 30 miles, as this gradient profile well illustrates.

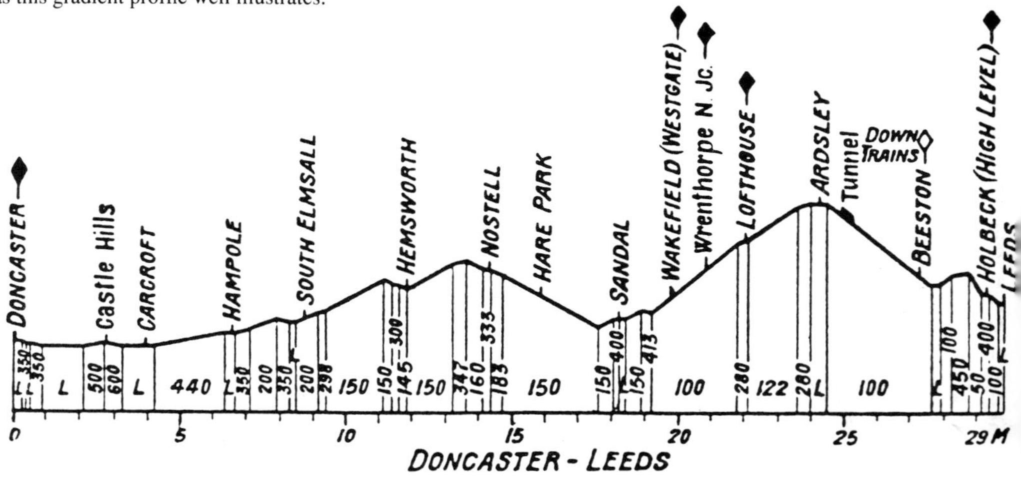

Above: A1 No. 60131 *Osprey* (10/48-10/65) takes an up relief express out of the platform at Wakefield Westgate c.1954. The locomotive is based at Leeds Copley Hill shed, then still part of the Eastern Region and coded 37B. This was changed to 56C in 1956 when the shed was transferred to the North Eastern Region during a reorganisation of Motive Power Depots in the Leeds area. *Osprey* spent its final years working out of Leeds Neville Hill shed. It was one of five A1s allocated there from July 1963 and often used as stand-by power for diesel diagrams. They were occasionally borrowed by Holbeck shed for duties over the Settle-Carlisle route to Glasgow. As with Doncaster (see previous view), Wakefield has changed much since this view was taken but the location is still recognisable, (Eric Treacy/Millbrook House Collection)

Right: Peppercorn A2 No. 60533 *Happy Knight* (4/48-6/63) has an easy task climbing the 1:100 through Beeston with the 5.10 pm Leeds-Doncaster train in 1960, running under express headlamps despite being made up of non-corridor stock. A steel panelled articulated pair – brake third/composite – leads the train. This was a regular duty for Doncaster-based Pacifics (No. 60533 was the only Eastern Region A2 after January 1950 and was resident at Doncaster between January 1960 and September 1962), usually an A1, although it was also used as a running-in turn. No. 60533 was fitted with a Kylchap double blastpipe and chimney and MLS multiple valve regulator in December 1949, one of five engines so equipped (see page 8), following the use of a double chimney on the last A2 No. 60539 *Bronzino*. (Eric Treacy/Millbrook House Collection)

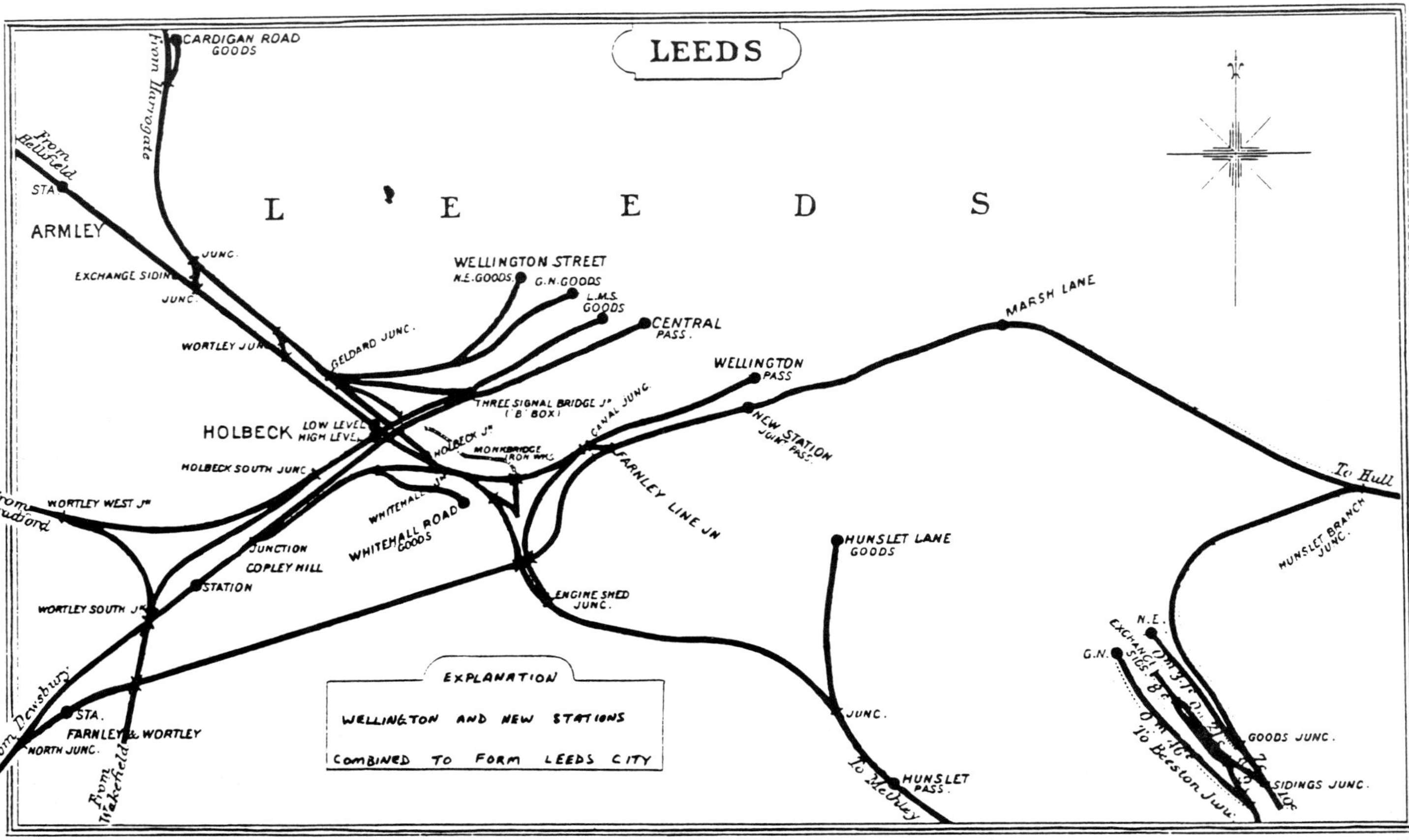

Above: The complex rail network in the Leeds area.

Right: Copley Hill based Class A1 No. 60118 *Archibald Sturrock* (11/48-10/65) leaves behind the industrial heart of West Yorkshire as it passes Beeston Junction on the climb out of Leeds with the up 'Yorkshire Pullman' in 1961-2. The train consists mostly of the new Metro-Cammell Pullman cars built in 1960-61 for the East Coast Pullman services. The design of these cars was based on BR Mk 1 stock and the cars were carried on Commonwealth pattern bogies. The leading coach is a second class kitchen car followed by a first class parlour car and a second class parlour car. One of the older, slab-sided, second class brakes brings up the rear (partly hidden by the bridge).

This collection of vehicles was the Harrogate portion of the train (which had reversed direction at Leeds) to which would be added the Bradford section at Wakefield and the Hull cars at Doncaster, the whole ensemble (usually by then between 10 and 12 cars) then going forward to London behind the same Leeds-based engine.

Ten North Eastern based A1s (No. 60118 was by then allocated to Neville Hill) were withdrawn in October 1965, following the end of the summer services, as the multiplication of the newer diesels (classes 37 and 47) severely cut the number of available duties, particularly south of York. (Eric Treacy/Millbrook House Collection)

Above: The up 'Queen of Scots' Pullman train passes Wortley South near Leeds hauled by A1 4-6-2 No. 60119 (11/48-5/64) in 1949. The Copley Hill based locomotive is in LNER green livery with British Railways in full on the tender. It received the name *Patrick Stirling* in July 1950 by which time it was painted in BR blue livery.

The 'Queen of Scots' was a through East Coast service from London to Edinburgh and Glasgow which diverted from the main line at Doncaster to serve Leeds, Harrogate and Ripon before resuming the 'direct' route at Northallerton. The trains 'reversed' at Leeds and two sets of eight cars sufficed for the full journey; but two extra cars worked between London and Leeds on the 'down' train, being then attached to the 'up' working for their return to Kings Cross. These are seen at the front in this view, thus making a formation of ten Pullman cars between Leeds and London. Note the LNER style headboard. The GNR line crosses over the original LNWR line between Leeds City and Farnley at this point (see map, page 22) and the signal gantry controls the junction for the GNR line to Bradford, avoiding Leeds Central. (Eric Treacy/Millbrook House Collection)

Left: The as-yet unnamed A1 No. 60119 again, this time blasting out of Leeds Central station with an up express c.1949. The engine is still in apple green livery and the teak-bodied Gresley coaches clearly have not received their new BR red and cream colours, so it is still very much an 'LNER' looking train.

Leeds Central was a relatively small terminus and any train of much more than eight coaches could scarcely be fitted into the platform length available. Prior to Nationalisation, it was also a 'shared' LNER/LMS facility (especially for ex-GNR and ex-L&YR services). Visual reminders of this can be seen on the middle road in the shape of GNR N1 0-6-2T No. 69437 (12/07-7/54) and on the right in the form of a Bank Hall (23A) based ex-LMS Black Five 4-6-0 No. 45216 waiting to take an express to Manchester and Liverpool via the L&Y route. (Eric Treacy/Millbrook House Collection)

Left: The lack of visible rivets on the tender is indicative of the Darlington origin of A1 No. 60136 *Alcazar* (11/48-5/63) of Grantham shed (35B) as it backs down on to a Kings Cross express at Leeds Central. Fairburn LMS-design 2-6-4 tank No. 42116 (1949-6/67) of Bradford Low Moor (ex-L&YR) shed (25F) is visible on the left of the picture. The date is probably around 1953 as there are no Derby lightweight DMUs in evidence; these pioneer units commenced operating the Leeds-Bradford services in June 1954. Note the miniature splashers necessary on the A1 design to accommodate the 6ft. 8in. driving wheels.

This view also gives a very clear impression of the distinctive front lamp irons of the Peppercorn 4-6-2s. Originally, the lower (wider) portion (see, for example, previous view) had carried permanent electric headlamps. But tradition died hard and they were always surmounted by conventional oil lamps anyway – so the whole exercise was rendered somewhat pointless as No. 60136 bears mute witness! As far as can be determined, the removal of the original lamps took place over a long period of time – see next view and others. (Eric Treacy/Millbrook House Collection)

Right: The coming of the larger and theoretically more powerful diesel classes spelt the end of Pacific domination over the East Coast main line. Initially the Class 40s intruded, but it was the delivery of the 22 Deltics and the later Class 46s which really robbed the 4-6-2s of their express work.

In this view taken at Leeds Central in the summer of 1964, Class A1 No. 60130 *Kestrel* (9/48-10/65), on a local train, is sandwiched between two Deltics working express diagrams. The steam locomotive still carries its original headlamps – see previous view – and is allocated to Copley Hill shed (56C). This depot closed in September 1964, so No. 60130 spent its last year at Ardsley (56B). The Deltic in the foreground, No. D9013 *The Black Watch* (9/61-12/81), is about to leave with the up 'White Rose' Pullman. This service was introduced in 1964 when the 'Queen of Scots' Pullman was discontinued north of Harrogate. The stock is mostly the new Metro-Cammell Pullman cars although the leading vehicle is an older brake (the 1960-1 batch contained no brake cars). Note how the eight-car formation occupies the full platform length and that, in consequence, the train engine is 'foul' of the carriage siding – a common occurrence at Leeds Central. In the platforms to the right are two Birmingham Railway Carriage & Wagon Co. DMUs (later Class 110), while behind D9013 is a Metro Cammell DMU (later Class 101). (Eric Treacy/Millbrook House Collection)

Right: The fireman of Class A1 No. 60141 *Abbotsford* (12/48-10/64) places the express code lamps on the lamp irons prior to taking charge of the up 'Queen of Scots' Pullman at Leeds Central c.1954. The engine is coupled to the two cars which remained at Leeds off the 'down' train (see page 23) and will draw them forward in due course to attach them to the outer end of the eight south bound cars which will arrive at Leeds from Glasgow via Harrogate and which will then reverse direction at Leeds.

The Copley Hill (37B) based locomotive has lost its electric lamps (see previous captions) but retains the steam generator and extended lamp irons. Note too the Automatic Train Control apparatus protected by a small metal coverplate; from the mid 1950s a larger plate with rounded bottom corners was used. (Eric Treacy/Millbrook House Collection)

Above: A 'foreigner' at Leeds (Holbeck) c.1963. A1 4-6-2 No. 60138 *Boswell* (12/48-10/65) stands outside the ex-Midland Railway shed in Leeds, which was transferred to the North Eastern Region in 1956 and was coded 55A; it had formerly been 20A. The locomotive, which was allocated to York (50A) all its life, has had its top lamp iron lowered still further by splitting the handrail on the smokebox door; note, however, that it still retains its original electric headlamps! It was probably being serviced at Holbeck after working a train into Leeds City, rather than working light engine out to Neville Hill. Former LMS Jubilee 4-6-0 No. 45589 *Gwalior,* a 55A engine, can be seen in the background. (Eric Treacy/Millbrook House Collection)

Right: Heaton (52B) based Thompson A2/3 No. 60517 *Ocean Swell* (11/46-11/62) approaches Holbeck Low Level with a Liverpool-Newcastle express around 1959. The Midland signal box in the background is Whitehall Junction, controlling the the Midland, North Eastern and old LNWR routes into Leeds City from the north and west. No. 60517 is on ex Midland metals at this point and will cross to the North Eastern line for Harrogate at Wortley Junction (see map page 22) The driver was probably displeased as the distant signal is at caution.

Like many through services via Harrogate, the Liverpool-Newcastle trains reversed direction at Leeds and in this case, also changed from ex-LMS to ex-LNER motive power. The A2/3s were a common sight on the Leeds-Harrogate-Ripon-Northallerton route; but an A2/3 was not the normal motive power on this particular service because this was usually a Heaton A3 diagram. The passageway in the foreground is the entrance to Holbeck Low Level station which closed in 1958. (Eric Treacy/Millbrook House Collection)

Left: York based A1 No. 60138 *Boswell* (12/48-10/65) passes Wortley Junction with the down 'Queen of Scots' Pullman c.1959. The train will have reversed at Leeds Central and two of the ten Pullman cars run from Kings Cross will have been detached. Though not by any means unique, the sight of an A1 on the 'Queen of Scots' north of Leeds was not commonplace. The normal motive power was a Gresley A3 and sometimes even Class V2s were pressed into service.

No. 60138 is already on the North Eastern tracks having come around the curve to Geldard Junction from Leeds Central, Holbeck Low Level station being just visible in the background. Note the larger AWS coverplate; this equipment was generally applied to the East Coast Pacifics from the late 1950s, following extensive trials on the Eastern Region. During daylight hours ordinary lamps were placed over the electric lamps on extended irons to give the necessary headcode; an early experiment was made with GER style discs, but these were soon replaced. (Eric Treacy/Millbrook House Collection)

Right: The up 'Queen of Scots' Pullman leaves Harrogate Platform 4 at 3.52 pm behind A2/2 No. 60501 *Cock o' the North* (9/44-2/60) in the mid-1950s. The appearance of a Class A2/2 on this working was somewhat rare, the normal motive power being a Gresley A3 at this time.

The photograph was taken from the road bridge above Harrogate South box which had a 45 lever frame. The down and through roads were signalled for bi-directional running, the signal visible is No. 20, the through road to down main starter. The NER clerestory brake is stabled in the up siding. The York (50A) based engine would come off the train at Leeds Central. Note the wooden scotch block at the entrance to the short siding on the left hand side activated from the box. Today (1990) the layout at Harrogate is vastly simplified. There is only one through road (hardly ever used as such) and the only long distance through trains are the (very) occasional HSTs to Kings Cross – usually no more than one per day. (Eric Treacy/Millbrook House Collection)

d) North Eastern Routes

Below: It is an old (but historically fairly accurate) story that in East Coast route terms, the GNR gave way to the NER in 'a field just north of Doncaster'. We have no precise view of this changeover point; but (resuming our survey of the ECML proper) there is no doubt that Selby marked the first significant point in North Eastern territory – and this is where we start.

In this view, A1 4-6-2 No. 60143 *Sir Walter Scott* (2/49-5/64) brings an up express over the River Ouse swing bridge and through Selby station in May 1953. This Gateshead (52A) based locomotive has received a lipped chimney and the electric lighting has already been removed. The signalling is still provided by NER pattern lower quadrants carried on wooden slot posts, although the gantry in the background has one upper quadrant replacement. An unusual feature of the swing bridge was the interlaced tracks on the down side which were replaced in 1960 when the control cabin was raised. Today, of course, East Coast main line expresses no longer pass through Selby, having been rerouted via the new diversionary line, opened in 1983 in consequence of the development of the Selby coalfield. (P. M. Alexander/Millbrook House Collection)

Above: The modified V2 series of Thompson A2s were a little camera-shy but here, Class A2/1 No. 60508 *Duke of Rothesay* (6/44-2/61) rounds the super-elevated curve at Barlby North Junction, Selby with an up relief express from York and the north on 4th August 1953. Note the New England based pacific's burnt smokebox door – a problem frequently encountered due to the poor performance of the asbestos rope seal between the smokebox and the door. The line from Hull trails in from the right. The track bed of the old main line has been converted into a road here, following the diversion of the East Coast main line, avoiding Selby, in 1983.

A noteworthy feature of the scene at this point well into BR days was the quite splendid gantry of lower quandrant NER signals, well seen here. In fact, it was often the infrastructure of the railway which revealed its origins long after the engines and carriages had changed to the new order. (John Robertson Collection/Colour Rail)

Right: Although the 'technical' change from GNR to NER territory was well south of the city, for all practical purposes, York was the true 'frontier' point. Here, we see a rather too obvious case of 'smoke by arrangement with the fireman', the latter checking to see if his efforts are appreciated!

Grantham A1 No. 60144 *Kings Courier* (3/49-4/63) is heading south out of York with a Kings Cross express in the mid 1950s and the particular photographer in question was renowned for obtaining 'footplate cooperation'! But York was often a notorious place from which to get a heavy train moving as is witnessed by the next view. (Eric Treacy/Millbrook House Collection)

Left: One of the authors was fortuitously on hand one Saturday afternoon in July 1949 when A1 No. 60138 completely 'lost her feet' under the old NER gantry with the up 'Scarborough Flyer'. The pyrotechnics continued for several minutes without any significant forward movement of the engine, and when the train had finally departed, the visible evidence remained on the railhead in the form of crescent-shaped depressions. It was normally northbound trains from Platform 9 (now Platform 5 – 1990 onwards) which gave rise to this sort of thing! For the record, No. 60138 was still in apple green livery at the time, had a stovepipe chimney and was un-named. It later became *Boswell*. (David Jenkinson)

Right: This almost identical view to the previous shot was taken at York in 1950 during the extensive resignalling programme which saw the replacement of the characteristic North Eastern Railway slot post semaphores by colour light signals. The new gantry is already in place behind the old NER version. By contrast, however, Class A1 No. 60158 (11/49-12/64), named *Aberdonian* in March 1951, eases an up afternoon express gently away from the platforms towards its home depot of Kings Cross (34A). At this time the locomotive was painted in early BR express passenger blue livery and had a 'proper' chimney! The worksplate is also displayed prominently on the cabside. To the right is the south end passenger pilot J71 No. 68246 (7/89-11/58) fitted with both steam and vacuum brake plus steam heating equipment. Just visible on the left is Class D49 No. 62710 *Lincolnshire* (2/28-10/60) and the similarity of composition tempts the thought that this too, may have been the up 'Scarborough Flyer' but without headboard. (S. E. Teasdale)

Left: A rather dirty Thompson standard 6ft 2in Class A2/3 pacific No. 60518 *Tehran* (12/46-11/62) stands at York Platform 9 with a down express in the mid 1950s. The 'SC' below the Gateshead shed plate denotes a self cleaning smokebox, but the driver on the platform casts a rather rueful look at the engine. The Thompson Pacifics were regarded as disappointing performers particularly after the A3s were fitted with double chimneys. Note the wheeltapper walking down the train at Platform 8 (now Platform 3 – 1990). (Eric Treacy/Millbrook House Collection)

Below: The first Peppercorn 6ft. 8in. A1 pacific No. 60114 *W. P. Allen* built at Doncaster in August 1948 waits to take a southbound express out of York c.1957 (note the mixture of carriage liveries – see page 14). No. 60114 was a Grantham engine at the time this picture was taken; however it moved to Doncaster (36A) in September 1957 where it remained until withdrawal in December 1964. The stock is all Mk I and includes the first BR Mk I catering vehicles, a full kitchen with supporting open vehicles, the fifth coach being an open first restaurant car. The standard series was produced from 1957.

By now, the magnificent NER signal gantries had long vanished – see previous views – but there was still much evidence of former company days, quite apart from the engine: the ex-LMS coaches on the right being quite typical of what might be seen. Today (1990), the magnificent sweep of the train shed roof remains unaltered but the track layout is greatly simplified in consequence of electrification and the two 'through' roads are no longer present. (Eric Treacy/Millbrook House Collection)

Right: This was the place where northbound steam trains could often encounter slipping problems at York – Platform 9 – but here, Thompson Class A2/3 *Watling Street* of Gateshead shed (52A) negotiates the crossings and sharp curves without undue concern as it leaves with the 1.00 pm Kings Cross-Newcastle express in the spring of 1949. This train was one of those selected to take part in the livery experiments carried out by BR in 1948 and the Thompson stock carries chocolate and cream livery with the numbers unusually placed centrally on the bodysides. The stock for the opposite working received plum and spilt milk livery. Although a number of Gateshead locomotives were painted in experimental blue with red and cream lining in order to work with these train sets, No. 60521 is still in LNER green and lettered 'BRITISH RAILWAYS' in full, for none of the Thompson or Peppercorn Pacific classes were included in the livery trials. On the right V2 No. 60979 (1/44-10/62) of York shed heads a local train. (Eric Treacy/Millbrook House Collection)

Below: Peppercorn A2 No. 60533 *Happy Knight* again, now allocated to New England shed (35A), leaves York Platform 9 with a down express on a fine summer afternoon in 1957. The train consists of Gresley, Thompson and BR Mk I stock in both maroon and red and cream livery. A Gresley articulated triple restaurant car set is seen at the platform end, the short central kitchen car section being just discernible behind the water crane. The building in the left hand background is the headquarters of the North Eastern Region, built by its predecessor the North Eastern Railway. The present day (1990) layout here is vastly simplified. (Eric Treacy/Millbrook House Collection)

Left: A1 No. 60154 *Bon Accord* heads the Royal Train near Eaglescliffe on 27th May 1960. The train was worked between Picton and Horden, north of Hartlepool, by the Gateshead based Pacific. The train had left Kings Cross the previous evening and had been stabled overnight at Picton. The extremely wet weather is necessitating considerable use of the sanding gear with this heavy set of vehicles. Identifiable carriages include the 12-wheel (ex-LMS) power brake No. 31209 (leading) followed by train staff dining car (ex-LNWR) No. 77, ex-LNWR Semi-Royal Saloon No. 806, ex-LNWR 12-wheel sleeping car No. 477 and ex-LNWR officers' saloon No. 45000. The rest of the train (including the Royal Saloons themselves) is out of the picture. The old LNWR portions of the Royal Train had a long and venerable life. Most of the vehicles illustrated lasted until the late 1960s, but the 12-wheel power brake and officers' saloon both continued in use until the end of the 1980s.

No. 60154 was one of five A1s fitted with roller bearings on all axles, noticeable in this view by the provision of only one mechanical lubricator (for the cylinders) on the left hand side of the running plate. (John Robertson Collection/Colour Rail).

Left: A Newcastle-Kings Cross express diverted via Leamside, County Durham, passes Wapping Bridge signalbox south of Penshaw headed by Gateshead Class A1 No. 60143 *Sir Walter Scott* (2/49-5/64) around 1957 or 1958. The up and down goods lines are on the right of the picture. (S. E. Teasdale)

Right: Class A1 No. 60127 *Wilson Worsdell* (5/49-6/65) heads a diverted up express through the disused station at Leamside, Co. Durham (closed 5th October 1953) on Thursday 14th May 1959 due to a derailment on the main line south of Plawsworth. In the 1959 summer timetable all Sunday through trains were allowed time for the Leamside and Washington diversion.

Diversions on the East Coast main line were common for a variety of weekend engineering works, but fortunately, all alternative routes were rated RA9 thus allowing through locomotive working. The letters 'RA' stood for 'Route Availability', the higher the number, the more restricted being the locomotive classes which could use the route in question. The 'RA' number was basically defined by static axle loading of the various locomotive classes. In consequence, the classification RA9 was a pre-requisite for any route which might be expected to 'accept' a high-powered engine. (S. E. Teasdale)

Below: A2/2 No. 60501 *Cock o' the North* leaves Sunderland with a Newcastle-Kings Cross express via the coast route to Northallerton calling at Sunderland, West Hartlepool, Stockton and Eaglescliffe, for a Middlesbrough connection and onto the ECML via Yarm in 1958/9. Note the later BR symbol on the tender and the AWS coverplate below the bufferbeam. (S. E. Teasdale)

CHAPTER THREE

NEWCASTLE, THE BORDER COUNTRY AND SCOTLAND

There is something about the Anglo-Scottish border country which even today, in the era of HSTs and 25Kv electrification, still places its mark upon the railway environment, whichever side of the country one might favour. Curiously enough, the transition seems to begin in all cases south of the border and in the case of the ECML, the changes, if not the frontier itself, were without doubt first hinted at in the fine city of Newcastle. Here, the railway abandoned its high speed characteristics and sterner stuff was now the order of the day!

For one thing, the arrangement at Newcastle itself revealed a different sort of station from those hitherto encountered – largely through accident of history – and the appended map shows something of its complexity in terms of approach tracks. It therefore seems only proper that before venturing further north, we should first take a reasonably detailed look at Tyneside itself, an area where the Thompson and Peppercorn pacifics somehow seemed to be very much 'at home'.

To understand the railway system at Newcastle, it is first necessary to appreciate that the though trains to Scotland, for many years, had to reverse direction at that city. They came in from the south via the High Level Bridge and then set out again in reverse direction via Manors and Jesmond for the north. It was only in the early days of the 20th Century (1906) that the opening of his namesake bridge by HM King Edward VII enabled significant changes to be made. But even though these much simplified the through operations at the station, they were often accompanied by fearsome curvature and highly complex trackwork: Newcastle was no place to rush through in unseemly haste.

Undoubtedly, and with very few exceptions, it was these factors combined with its distance from London which made Newcastle (like Carlisle on the West Coast line) a significant changeover point for locomotives no matter how much layout rationalisation might have taken place. By the time they had reached here from the south, engines would probably have a choked grate and be ready for relief – and this remained true for the whole of the steam era. The replacement engine may well be of the same design as the one which came off but it was not quite the same as a modern HST which only needs a change of driver. And so it was that during those final years of steam, Newcastle was to remain one of the last bastions of traditional main line express locomotive operation.

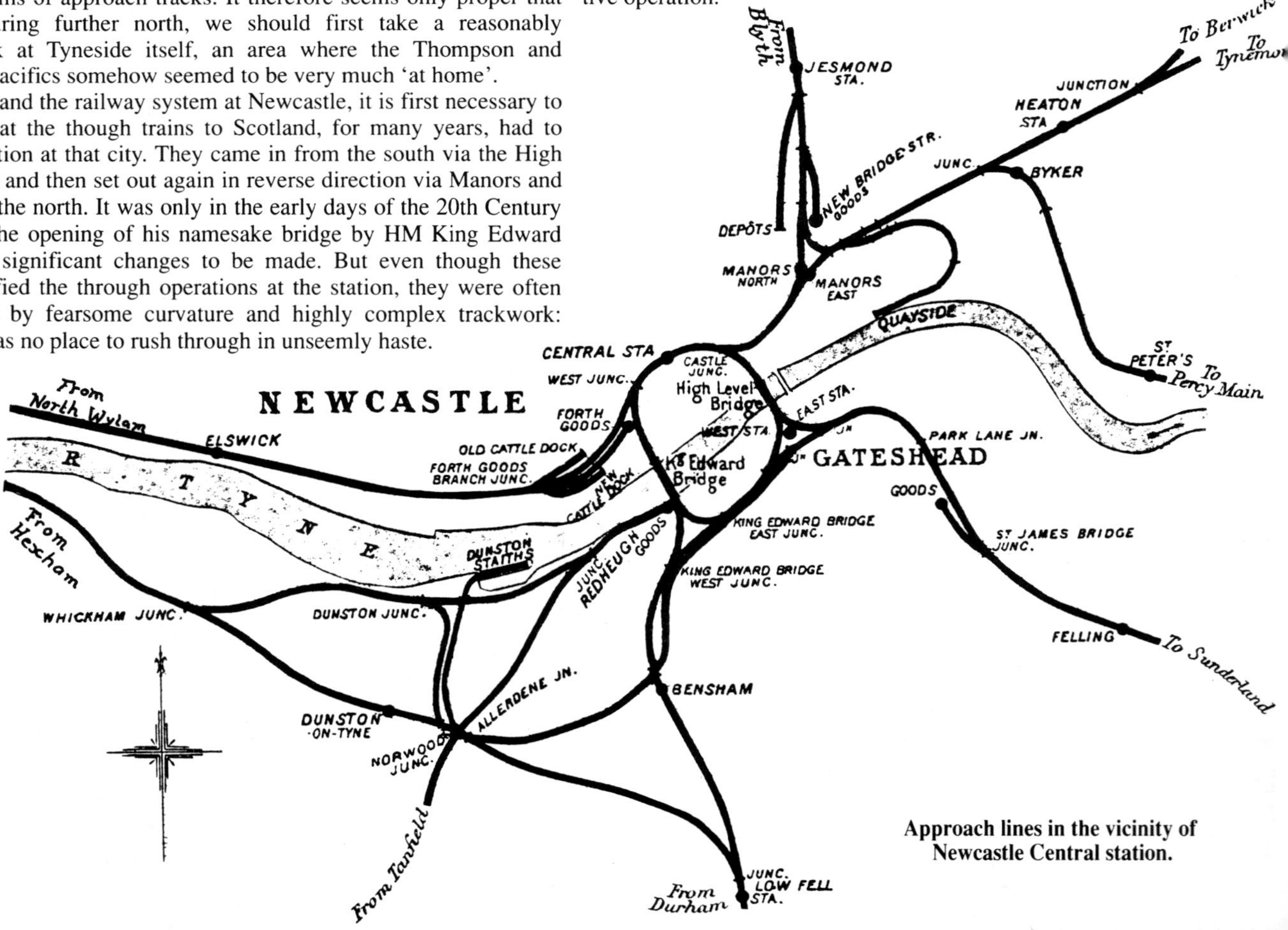

Approach lines in the vicinity of Newcastle Central station.

a) Tyneside

Right: Gateshead A1 No. 60150 *Willbrook* (6/49-10/64) heads the up 'Flying Scotsman' through King Edward Bridge West Junction c.1953-4 as it comes off the tight curve and gathers speed for its south bound journey. The signals are still largely North Eastern Railway lower quadrants, operated electro-pneumatically. Another NER survival is the ½mile post to the left of the picture. The headboard was often placed on the bufferbeam lamp iron as eddy currents caused the exhaust to obscure the driver's view when it was placed on the top lamp iron, it also happened to be far easier to put it there! The train itself is also interesting. By now, the BR Mk I carriage had become well-established, but the famous 'Flying Scotsman' was still comprised of post-war LNER Thompson stock specially built for this service and identified by the lower side 'fairings', partly concealing the solebars of the carriages. (Eric Treacy/Millbrook House Collection)

Left: The 2000th locomotive built at Doncaster works, was the pioneer Class A2/3 No. 60500 *Edward Thompson* (5/46-6/63) itself, here seen bringing the 10.10 am Edinburgh-London express over the King Edward bridge at Gateshead c.1953. The bridge was formally opened by King Edward VII on 10th July 1906 and opened to traffic on 1st October 1906. The check-railed nature of the tracks reveals all too clearly the price which had to be paid to avoid reversal at Newcastle.

The service depicted left Newcastle at 12.46 pm and arrived back at No. 60500's home town of Peterborough at 4.44 pm. This view also gives a good impression of the as-built quasi-standard Thompson A2 type (sub-classified A2/3) with its stovepipe chimney, 'round' dome cover and 'full' smoke deflectors, the latter having but a small 'cut-away' at the rear to facilitate access to the footplate top. (Eric Treacy/Millbrook House Collection)

Above: York based (50A) A1 No. 60138 *Boswell* (12/48-10/65) crosses into County Durham over the King Edward bridge at Newcastle with the up'Queen of Scots' c.1959. The locomotive would work the train via Harrogate to Leeds, where in all probability another A1, allocated to Copley Hill, would take the train forward to Kings Cross on a lodging turn. The train is the customary 8-car formation which formed the main through portion of the working between London and Glasgow – see page 23. Again, the check-rails (denoting the severe curvature) may be noted, starting at the very end of the bridge. (Eric Treacy/Millbrook House Collection)

Below: This shot from the footplate of an immaculately turned out Class A1 No. 60135 *Madge Wildfire* (11/48-11/62), a Gateshead (52A) based locomotive, taken as it crosses King Edward bridge with an up express, again gives a vivid impression of the track geometry at Newcastle. (Eric Treacy/Millbrook House Collection)

Right: This superbly crisp view shows Darlington built A1 No. 60147, appropriately named *North Eastern* (4/49-8/64), of Gateshead shed (52A) pulling away from Platform 8 at Newcastle with the up 'Flying Scotsman' c.1953. The visible stock is all Thompson designed: a full brake and corridor first are followed by three catering vehicles, similar in make up to the pre-war Gresley sets except that they were not articulated. An open first restaurant car leads, next is a full kitchen car and out of sight (behind A1 No. 60138 *Boswell* on the right) is an open third restaurant. The A1s were often used on the 'Flying Scotsman' but not on the non-stop Kings Cross-Edinburgh services as corridor tenders were never attached ot this class. Their steam brakes made the A1s incompatible with the vacuum fitted corridor tenders. (Eric Treacy/ Millbrook House Collection)

Right: The crew of former Class P2 2-8-2 *Thane of Fife* (ex-No. 2005, rebuilt 1/43-11/590 have ensured they will have sufficient coal for their journey south with the 12.46 pm departure from Platform 8 at Newcastle, the 10.10 ex-Edinburgh for Kings Cross around 1957 – but note the retained stovepipe chimney. The 9 ton capacity tender is absolutely full after servicing at Gateshead – much of the extra coal will probably not get to the firebox! The New England based locomotive would probably have worked north from Peterborough with a Kings Cross-Edinburgh sleeping car train in the early hours. No. 60505 was one of two A2/2s to be withdrawn in November 1959, the first of the former LNER Pacifics to be condemned, other than the war damaged A4 No. 4469. (Eric Treacy/Millbrook House Collection)

Below: A1 No. 60124 *Kenilworth* (3/49-3/66) of Gateshead shed (52A) eases a northbound express, probably the 9.15 am Leeds-Glasgow, over the celebrated diamond crossings at the east end of Newcastle Central station c.1959. A J39 0-6-0 heads a short freight on the goods lines to the left of the picture, a J72 0-6-0 T stands as station pilot to the right and a Tyneside electric train is visible in the platforms. Prior to the building of the King Edward Bridge in 1906, northbound trains would enter Newcastle (prior to reversal) via the lines in the left foreground. (Eric Treacy/Millbrook House Collection)

b) The East Coast Main Line North from Newcastle an appraisal *by A. J. Powell*

The northern end of the East Coast main line, and its continuation from Edinburgh to Aberdeen linked a number of 'frontier' cities. Apart from the non-stops (worked by Kings Cross and Haymarket men and A4s) and the Leeds-Glasgow 'North Briton' (usually an A3 job), engines were religiously changed at Newcastle. Workwise, Gateshead depot went south (including the lucrative London lodging turns) and Heaton went north; any trespass was, to say the least, discouraged. Even then, neither engine or men worked *through* Edinburgh; Haymarket worked the expresses north and south, St. Margarets the main line freight, and *everyone* knew their place – although the Edinburgh depots would 'borrow' an engine if short of motive power. Likewise, on the Aberdeen road, everything changed at Dundee Tay Bridge; in this city on the Firth of Tay, the water supply at the station was unreliable and it was quicker to change engines (and men) than to fill the tender tank.

Another point of commonality was that while high speeds might be all right further south, in Northumbria and Scotland they were not to be

Above: The north end of Newcastle station with the up 'Heart of Midlothian' approaching under No. 1 signalbox (the former electro-pneumatic box closed after colour-light resignalling in 1959) hauled by class A1 No. 60161 *North British* (12/49-10/63) in the mid 1950s. The afternoon services between Kings Cross and Edinburgh were named in 1951 and given rakes of BR standard stock introduced that year. To the right of No. 60161 is Gresley V3 2-6-2 tank No. 67654 (3/36-9/63) on pilot duty and on the left, just visible across the platforms, can be seen a train of Gresley designed Tyneside electric stock. (Eric Treacy/Millbrook House Collection)

Left: The sole Scottish based Thompson Class A2/3 No. 60519 *Honeyway* (2/47-12/62) was allocated to Haymarket depot (64B) where it was to spend all but the last year of its life though rarely seen on the Aberdeen road – see text. It is seen here in Newcastle station at Platform 8 in the early 1950s about to head north. The engine still displays its original stovepipe chimney and dome and we are unable to offer an explanation for the 'reversed' head board above the buffer plank. An A8 4-6-2 tank can be seen across the platform with a local train. Note the swan neck water crane above the tender, altered in order to fill the high sided tenders of the LNER Pacifics. (Eric Treacy/Millbrook House Collection)

expected or encouraged. There was too much curvature for anything much more than 70 mph save in a few areas. Even the prewar streamlined 'Coronation' train was allowed a level two hours for the 124.4 miles between Newcastle and Edinburgh, and that was reckoned about as fast as passenger comfort would allow.

Then there were the slacks to 35 mph round the Morpeth curve (the place which later gave its name to the BR principle of installing AWS magnets on the approach to permanent speed restrictions), to 60 at Alnmouth (awkwardly placed for the four mile climb at 1:170 over the Little Mill hump) and to 40 over the Royal Border bridge followed by 30 through Berwick station. At least the slow climb from there to Burnmouth gave time to appreciate the magnificent cliff scenery! There was also the tortuous climb up the valley of the Eye Water as the line crossed and recrossed the burn – which took out seven bridges in its 1948 rampage. And just when the East Lothian coastal plain gave some scope for fast running, the avoiding line at Dunbar demanded a reduction to 60 mph. Drivers of up expresses would have relished another 15 mph on that stretch, so that Cockburnspath bank, 4½ miles of 1:96, could have been rushed.

So there was a constantly changing demand for power and careful control of the boiler was necessary to avoid blowing off. From the locomotive standpoint it was a waterless waste with only the oasis of Lucker troughs to relieve it – and even they were a wearisome 74 miles from Edinburgh. Then there was a plethora of level crossings, which in the 1950s increasingly brought a risk of distant signals 'on'. In these circumstances the A1s with their 6ft 8in wheels had no speed advantage over the smaller wheeled A2s. Indeed it could be argued that the A1s and most of the A2s, with their big 50 square foot grates were less suitable machines for the line than the A3s, A4s and V2s. Their great power was a bonus on one or two of the banks of course, but they tended to be heavier on coal and water than the earlier engines.

But if the line south of Edinburgh had its operating difficulties, they paled beside those of the Aberdeen road. For the locomotive it was a stinker. It had been established by joining up odd bits of local railway, much of it on difficult grades, and this ancestry was very apparent. From Saughton Junction, three miles out of Edinburgh, the 1890 Forth Bridge lines led to Inverkeithing, and included a vicious 1:70 climb to the bridge from the north. The 25 mph speed restriction through Inverkeithing reflected the main line joining the original North Queensferry branch and promptly diverging from it to head for Burntisland via a summit at Dalgety. There the line was on a 25 mph curve to join the original alignment from the ferry pier. The excavators did not meet accurately in Kingham tunnel, resulting in a 30mph restriction over some remarkable convolutions. Thornton Junction was a permanent mining subsidence area – the engineers had forgotten how many times they had built up the island platform – and needless to say it was at the foot of hard climbs in both directions.

Further north, at Ladybank, the route parted company from the original straight line to Perth, but got no further than Leuchars before curving away from the Tayport line to make for the severely restricted Tay Bridge. Once across the Tay and through Dundee station and Dock Street tunnel, the route sidestepped to join the level Dundee and Arbroath Joint at Camperdown Junction. Now at long last some speed was possible. But from Arbroath station the line turned north on a checkrailed curve on the original branch to Forfar, only to discard it after ⅓ mile at St. Vinean's Junction. The next impediment was the two mile single line section from Usan box down into Montrose, requiring a dead slow to pick up and drop the token; then at Kinnaber Junction, Edinburgh trains were severely restricted in joining the Caledonian line to Aberdeen, on which at last, despite some taxing gradients, running could be fairly unrestrained until the final sweeping curves into the Granite City. And the LNER regarded this little lot as a main line!

Before the war Gresley had realised that something special was needed to work the line, but only scored a near miss with his Class P2 2-8-2s.

Even the final Thompson A2/3s with their troublesome, whippy front end, were no better match for the curves and No. 60519 was discreetly kept off the Aberdeen road as far as possible. Which left Haymarket's small fleet of five Peppercorn A1s along with five Peppercorn A2s at Dundee and Ferryhill. Their duties included not only the express passenger service but also the Kings Cross meat and fish trains and the overnight fitted freights. Hardly sufficient and if something geared specifically to the nature of the line was needed, perhaps Peppercorn would have been better advised to mount an A4 boiler with double chimney on an A2 chassis for the best of both worlds.

The other line to see the A1s and A2s alongside V2s was that to Perth, branching off at Inverkeithing; its nature was not much better than the Aberdeen road, though mercifully it was only 35 miles long. Northbound the main work was on the 7½ miles of hard climbing past Dunfermline to Cowdenbeath, mainly at 1:89/74/100, and southbound in getting from the valleys of the Tay and Earn up the 6½ miles of 1:75 to Glenfarg – now mostly turned into a motorway route! Between the two banks there was an extremely broken profile, but curvature was reasonable; the stumbling block was mining subsidence round Cowdenbeath and in the Kelty area. The passenger traffic was not substantial, but there was a modicum of through express freight for the Haymarket Pacifics to share.

One of the surprising results of nationalisation was the transfer of three A1s to Polmadie in 1951/2 to work on the West Coast main line (see page 58). They established a regular routine of working Glasgow-Birmingham trains as far south as Crewe; on this road their sustained boiler power was much more matched to requirements, and they were generally liked by enginemen. They were reckoned a little lighter on coal and water than the usual 'Duchesses' and 'Princesses'.

In their last years they began to wander onto new routes as diesels displaced them from their usual haunts. No less than ten Heaton engines went off to Tweedmouth after the 1962 summer timetable for freight work or storage. A year later five Haymarket A1s went to St. Margarets, also for freight work, some of it over the Waverley line. It was a last despairing effort from one time top link passenger engines.

AJP

The convoluted crossing of the Firth of Tay in the Dundee area.

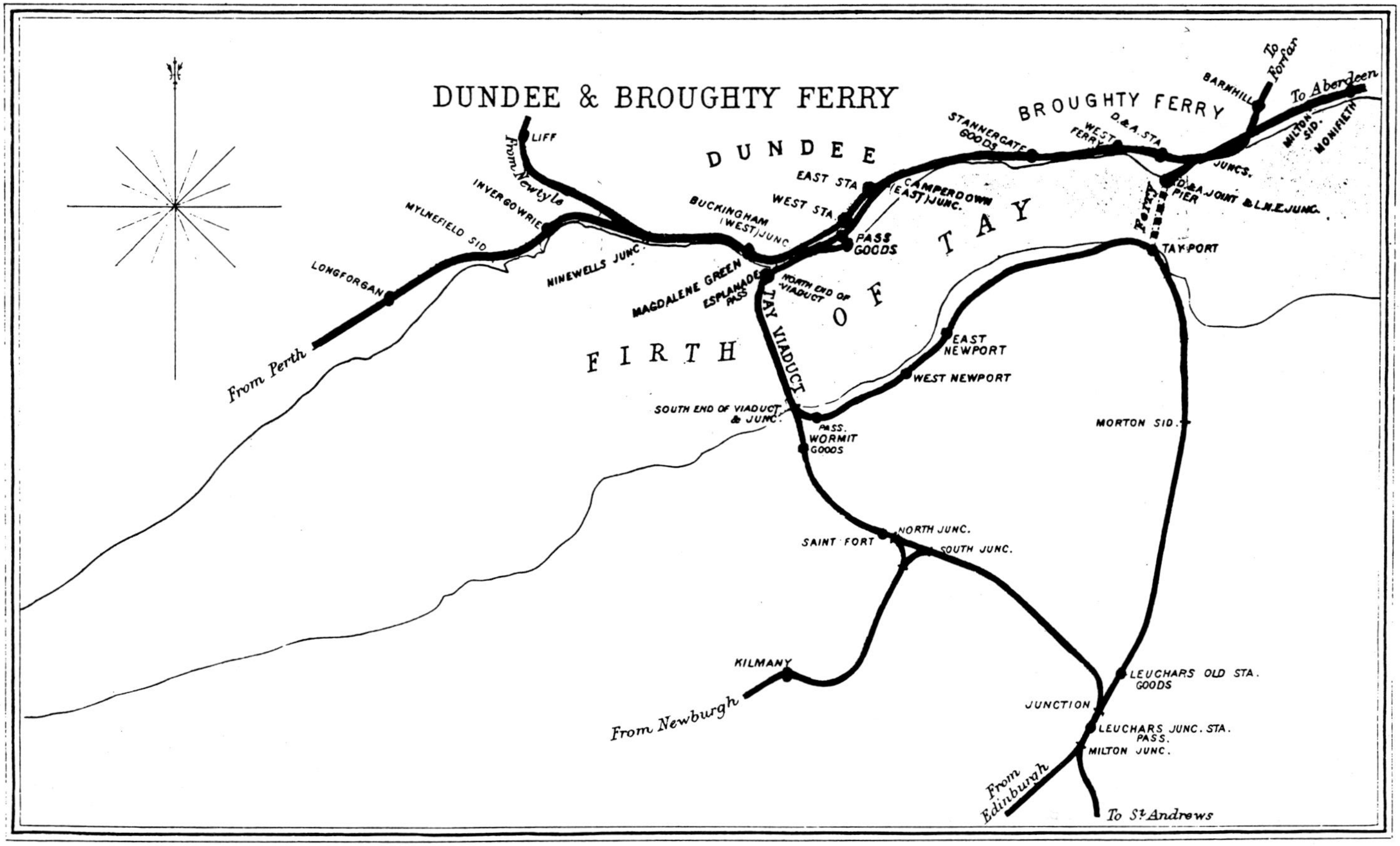

c) Berwick and Cockburnspath

Left: Thompson A2/3 No. 60517 *Ocean Swell* (11/46-11/62) of Heaton shed (52B) pulls away from Berwick upon Tweed with a southbound express in the mid 1950s. Of almost equal interest is the considerable array of pre-group lower quadrant signals and the well-high 'pure LNER' composition of the set of carriages on the left. (Eric Treacy/Millbrook House Collection)

Left: Peppercorn Class A2 No. 60536 *Trimbush* (5/48-12/62) tackles the 4½ miles of 1:190 north of Berwick with a down express freight in July or August 1955. The train includes a variety of wooden and steel bodied wagons including a few bottom door types. No. 60536 was allocated to Haymarket (64B) throughout the 1950s, but ended its days at St. Margarets (64A). It may have been running in after overhaul or even making up mileage prior to shopping, but this picture reveals that well before the final onslaught upon the steam locomotive, even the best of the breed could be seen on quite humdrum duties. (Eric Treacy/Millbrook House Collection)

Above: Peppercorn A2 No. 60534 *Irish Elegance* (4/48-12/62) of Haymarket shed (64B) rounds the curve out of Penmanshiel tunnel, still climbing the last stretch of Cockburnspath bank with the up 'Heart of Midlothian' on 1st September 1956. The train left Edinburgh at 2.00 pm and contained through coaches from Aberdeen, which were also likely to have been A2 hauled to the Scottish capital. Penmanshiel Tunnel is, of course, no longer with us, having been closed following the disastrous landslip in 1979 which caused the subsequent diversion of the East Coast Main line, leaving the tunnel itself as a sombre but reverent tomb and memorial to those brave men who lost their lives in that frightful disaster. (John Robertson/Colour Rail)

Right: North Eastern Region Thompson Pacific A2/3 No. 60521 *Watling Street* (5/47-11/62) of Gateshead shed (52A) climbs Cockburnspath bank with a down parcels train in the mid 1950s. Note the new bridge, one of the replacements necessary after the extensive flood damage of 1948. The train is an interesting mixture of 4-wheel, 6-wheel and bogie vans with two ex-LMS 50ft full brakes leading. The first is a Stanier design followed by a wooden panelled example which has recently been refurbished with steel panels below the waist. The locomotive itself spent its last year at Tweedmouth shed, displaced from express duties by dieselisation – in particular the introduction of the Deltics in 1961/2. (Eric Treacy/Millbrook House Collection)

Left: Class A2/3 No. 60516 *Hycilla* (11/46-11/62) hauls a down parcels at Monktonhall Junction near Edinburgh in August 1958. The Gateshead (52A) based engine has been fitted with AWS equipment and still carries the Thompson round dome boiler (Diagram 117) together with a plain double chimney. No. 60516 ran some 833,050 miles before her withdrawal from York shed (50A) in November 1962 and this view is typical of the type of work to which these engines were allocated in their final years. (D. Cross)

d) The Waverley Route

Above: This later period view of Thompson Class A2/3 No. 60519 *Honeyway* shows the engine passing under Portobello East signalbox in the summer of 1958 or 1959 with an up express. This train is taking the route for Newcastle, but Portobello East was the junction for the celebrated Waverley route, where southbound trains turned right off the ECML and headed through the 'Walter Scott' country for Carlisle. The beaded stovepipe chimney has been retained but other changes have occurred since the mid-1950s: the numberplate and top lamp iron have been lowered, the later BR symbol has been applied and more significantly, an A1 boiler with a banjo dome has replaced the round dome Thompson variety. (Eric Treacy/Millbrook House Collection)

Left: Peppercorn A2 No. 60527 *Sun Chariot* (1/48-4/65) comes down the 1:75 from Whitrope summit on the Waverley route near Shankend with a Carlisle-Edinburgh express around 1959. The shedplate reads 62B Dundee and the locomotive may have been 'borrowed' by Haymarket! This was one of three Peppercorn A2s (Nos. 60530 and 60535 were the others) allocated to Polmadie (66A) fromSeptember 1963 to replace the LMS Pacifics then being withdrawn. They were used on WCML duties but not surprisingly, proved unpopular with the Polmadie crews! Even so, Nos. 60527 and 60535 remained there until withdrawal. (Eric Treacy/Millbrook House Collection)

Above: This picture is an early (c.1949-1950) view of the pioneer example of Thompson's Pacific version of the V2 2-6-2 No. 60507. These engines (Classified A2/1) first appeared in May 1944 when No. 3696 emerged from Darlington works. Under the revised 1946 LNER renumbering scheme, this locomotive became No. 507 (ultimately BR No. 60507) and was named *Highland Chieftain* in May 1947. No. 60507 moved to the Scottish Region's Haymarket depot (64B) in December 1949 and was withdrawn from St. Margarets in December 1960.

The 'Scottish' naming theme (also adopted with the other three examples, Nos. 60508-10) was conceivably devised as a follow-on to the names of the former P2 engines (page 4) but whether V2 derived, or ex-P2, the A2/1s and A2/2s were not a conspicuous success. It is worth recalling however that the V2 derivatives (BR Nos. 60507-10) carried the full size smoke deflectors of the 'proper' Thompson A2/3 series, whereas the rebuilt P2s were always distinguished by their lack of full deflector shields.

The location shown here is Port Carlisle Branch Junction where trains for the North British line to Edinburgh left the Caledonian main line. *Highland Chieftain* is in charge of the mid-morning London (St. Pancras) to Edinburgh (Waverley) express comprising LMS stock, many vehicles still in crimson lake LMS livery. The third coach is a BR-built but LMS-designed porthole composite in an early version of red and cream livery with the number at the left hand end of the bodyside. This type (not built until after Nationalisation) was an 'all-steel' derivation of the traditional Stanier coach and formed an important intermediate step towards the BR Mk. I design. Catering facilities on this train, however, are still provided by the traditional ex-LMS pre-war twelve wheel dining car, fifth in the rake. (Eric Treacy/Millbrook House Collection)

Below: Carlisle 1960 with Peppercorn A2 No. 60535 *Hornets Beauty* (5/48-6/65) leaving Citadel station with a mid afternoon four coach local for the Waverley route calling at most stations to Edinburgh. Note the recently fitted Smith Stone speed indicator. This Edinburgh Haymarket (64B) engine moved to Edinburgh St. Margarets in October 1961 as diesels took over the East Coast main line work and moved yet again in September 1963 to the Caledonian depot at Polmadie (66A) where it took over some of the duties of the 'native' Stanier locomotives. (Eric Treacy/Millbrook House Collection)

Below: The familiar roofline of Carlisle Citadel station frames Class A2/3 No. 60519 *Honeyway* as it moves away from Platform 4 with a mid afternoon train for Edinburgh (via the Waverley Route) circa 1955. The train shed is looking somewhat dilapidated and was rebuilt around 1957. However the elegant and functional combination of two different styles of Gothic arch can still be admired in this view. As noted elsewhere, No. 60519 was the only A2/3 based north of the border and as can be seen from this and other views, it operated in a very wide parish! Note the original boiler and chimney and (in the left hand platform) a neat LMS non-corridor set of c.1931 vintage. (Eric Treacy/Millbrook House Collection)

e) Edinburgh (Waverley and Haymarket)

Right: All the Peppercorn A1s appeared without names, by contrast with the A2s (of all variants) which had them from new. No. 60152, the last Pacific built at Darlington works (7/49-6/65), ran for almost two years before receiving the name *Holyrood* and was repainted BR blue at the same date, June 1951, a matter of weeks before this livery was abandoned in favour of green. This Haymarket based engine is seen here leaving Edinburgh Waverley in sparkling apple green livery and early BR markings, with an up express around 1950. The train is not identified but such carriages as can be discerned are of the latest-LNER pattern (many with lower side fairings and all but one in 'teak' livery). This and the obvious 'newness' of the engine infers an important working. (C. C. B. Herbert Collection/NRM)

Right: Former Class P2 2-8-2 No. 2005, now Class A2/2 No. 60505 *Thane of Fife* (1/43-11/59) and based at New England (Peterborough) takes the up 10.10 am relief 'Scotsman' out of platform 10 at Edinburgh Waverley c.1952-3. The leading coaches are still in LNER teak livery, including the NER designed brake composite which heads the rake. It is unclear why No. 60505 should have ventured as far north as Edinburgh (New England lodging turns usually worked to Newcastle only), so the engine may have been borrowed by Gateshead to replace a failure, thus finding it way to Edinburgh, undoubtedly familiar territory when it was a 2-8-2. (Eric Treacy/ Millbrook House Collection)

Above: This view at Edinburgh Waverley (west end) shows A1 No. 60160 *Auld Reekie* (12/49-12/63) moving out of platform 11 with an afternoon express for Glasgow or the north in the mid 1950s. This Haymarket based locomotive, carrying the appropriate Scottish 'nickname' for Edinburgh, was one of three which also worked out of the former LMS depot (ex-CR) at Polmadie, Glasgow (66A) between 1951 and 1953, No. 60160 returned to Haymarket early in 1952. (Eric Treacy/Millbrook House Collection)

Right: This splendid semi-formal portrait shows North Eastern Region class A1 pacific No. 60132 *Marmion* (10/48-6/65) of Gateshead shed (52A) standing by the mechanical coaling plant at Edinburgh Haymarket depot on 12th October 1952. The engine was one of 23 Class A1s to be constructed at Darlington, instantly recognisable by the smooth tender and cabsides by contrast with the Doncaster-built locomotives which had snap head rivets on these surfaces. Note the men in the wagon half way up the coaling plant! To the right is a water tower constructed from prefabricated sections. (John Robertson/Colour Rail)

Left: Haymarket (64B) based A2 No. 60537 *Bachelors Button* (6/48-12/62) waits at the east end of the shed, whilst former LMS Black 5 No. 45361 from Inverness shed (60A) is turned on the vacuum operated 70ft turntable. Note the white rim around the perimeter of the table. The wording on the notice to the left of the Class 5 reads: "Both locks must be in before engines are allowed to move on to or off this table". Further left is a man carrying signal lamps.

No. 60537 was one of the locomotives moved to Scotland in 1949 to replace the Thompson A2/2s on the Edinburgh-Aberdeen main line and it is also worth noting (perhaps irreverently) that the chosen name was hardly appropriate to such a noble looking locomotive! This was quite a characteristic consequence of the LNER-inspired 'racehorse' naming theme and was by no means confined to the A2s. Some of the A3s had equally odd-sounding names – *Pretty Polly,* for example! The date of this picture is probably c.1955-6, as the top lamp iron and numberplate have been lowered, a modification completed in March 1956 on the A2s, but before the introduction of the later BR symbol. The original electric headlamps, however, remain 'in situ'. (Eric Treacy/Millbrook House Collection)

Left: Peppercorn A2 No. 60535 *Hornets Beauty* (5/48-6/65) begins to sparkle as it receives attention from the Haymarket cleaners, whilst Heaton (52B) based K3 2-6-0 No. 61987 (1/37-3/62) retains its covering of grime – the date is 1957. The six Haymarket Peppercorn A2s worked to Newcastle, Carlisle, Perth and Glasgow as well as the Dundee and Aberdeen route. Both the locomotives in the picture are three cylinder types – the Gresley design on the right with derived motion for the middle cylinder: note the 2:1 lever access hatch. Whilst Peppercorn followed Thompson in providing three separate sets of Walschaerts valve gear, one cannot help but comment upon the remarkable similarity of 'face' between the Peppercorn and Gresley designs – truly LNER and quite distinctive to boot. (Eric Treacy/Millbrook House Collection)

f) The Great Bridges and the Aberdeen Route

Right: This unusual and striking view, taken above the track of the Forth Bridge, shows Peppercorn A1 No. 60160 *Auld Reekie* again (12/49-12/63), this time hauling an up meat train. The date is probably around August-September 1957 just after No. 60160 received a Thompson Diagram 117 boiler, with false banjo dome cover. Note the white refrigerated containers carried in open wagons. (Eric Treacy/Millbrook House Collection)

Above: Originally ordered as a V2 2-6-2 (see page 3), Thompson Class A2/1 4-6-2 No. 60509 *Waverley* (11/44-8/60) of Haymarket shed runs through Dalmeny station with an Aberdeen-Edinburgh express in the mid 1950s. Until Nationalisation in 1948, this station had been owned by the Forth Bridge Railway Company, formed by a consortium of pre grouping companies, which was also responsible for the Forth Bridge itself.* The basic railway structures are, however, entirely North British in character. (Eric Treacy/Millbrook House Collection)

* Footnote: For a fuller account of the history of the Forth Bridge Railway, see 'Rails above the Forth', published in Atlantic Transport Publishers' 'BACKTRACK' Volume 4, Nos. 2 and 3 (1990).

Left: This peaceful view shows one of the Scottish based Class A1s, all based at Haymarket (64B) No. 60159 *Bonnie Dundee* (11/49-10/63) running through North Queensferry with an Edinburgh Waverley express on 15th June 1957 – note the later (1956-7) BR symbol on the tender side. The station displays an array of holiday posters (including one advertising the attractions of Burnham on Sea, Somerset) for the more adventurous Scottish traveller and it is hard to believe that this picture is scarcely more than a generation old (written in 1990). (John Robertson Collection/Colour Rail)

Right: No. E529 *Pearl Diver* (2/48-12/62), one of the earlier Peppercorn A2s, heads south from Tay Bridge with a fitted express freight carrying meat in the winter of 1948-9. At this time No. E529 was the only Peppercorn 4-6-2 allocated to the Scottish Region and its first depot was Edinburgh Haymarket where it remained until October 1961. The livery is LNER green as applied to No. 525 but by February 1948 the first signs of Nationalisation were appearing. The tender has 'BRITISH RAILWAYS' in full on the side and the number has the prefix 'E' above it on both the buffer plank and cabside. This treatment was given to Nos. E527-531 before the five figure BR numbers were used. The engine must have been considered a success on the Aberdeen line, for by the beginning of 1950, ten of her sister engines had joined her in Scotland. The engine was fitted with a double chimney and multiple valve regulator in September 1949 when it also received its BR number (60529) and was repainted BR green. (Gavin L. Wilson/David Jenkinson Collection)

Left: The last of the former Gresley Class P2 2-8-2s to be rebuilt was No. 2003 *Lord President,* emerging from Doncaster works in December 1944 as a class A2 Pacific. It is seen here, now reclassified A2/2 and renumbered 503, with the 12.30 pm Aberdeen-Kings Cross fish train (running under headlamps denoting that the train is classed as 'coaching' stock) near Saint Fort on 19th May 1948. The locomotive is still in LNER post war green livery with the allocation 'Haymarket' on the bufferbeam, although it was at the time allocated to Aberdeen Ferryhill for a period of only ten days! Note how the smoke is hugging the top of the boiler demonstrating the relative ineffectiveness of the small smoke lifters. As BR No. 60503, the engine ended its days south of the border at York shed (50A) from whence it was withdrawn in November 1959. (Gavin L. Wilson/David Jenkinson Collection)

Above: A view of Class A2/1 No. 60510 *Robert the Bruce* (1/45-11/60) during 1948/9 heading a down express approaching the south end of the Tay Bridge near Saint Fort. This engine, one of four originally conceived as V2 2-6-2s, was named in April 1948 when it received large smoke deflectors. Initially, small wing plates similar to the A2/2 deflectors were carried by the class and it still retains the six wheel style of tender (appropriate to the V2) which was replaced by an eight wheel version in September 1949. The Haymarket based engine is in plain black livery with 'BRITISH RAILWAYS' in full on the tender side. Note the small electric lamp fittings and the generator mounted on the rear bogie axle. The train is of pure LNER origin (some of the vehicles are actually pre-group) and in LNER livery. (Gavin L. Wilson/David Jenkinson Collection)

Right: Haymarket based A2/2 No. 60504 *Mons Meg* (rebuilt ex-Class P2 2-8-2 No. 2004, 11/44-1/61) comes off the Tay Bridge with an Aberdeen-Edinburgh express around July 1949. The leading coach is a Gresley tourist stock buffet car No. E9141 – built by the Birmingham Railway Carriage & Wagon Co. in 1935 using plywood panelling. The livery is the wartime/post war utility brown, not the original pre-war green and cream, and the carriage was probably attached with the locomotive at Dundee, hence its highly inconvenient position in the train! The six P2 rebuilds were all initially returned to Scotland but proved disappointing. When sufficient Peppercorn Pacifics became available the A2/2s moved south to New England and York, No. 60504 being the last to leave Scotland in January 1950. (Gavin L. Wilson/David Jenkinson Collection)

Dundee-based class A2 No. 60527 *Sun Chariot* (1/48-4/65) moves out of the bay at Dundee Tay Bridge station to back onto an express for Edinburgh in 1949/50. Interestingly, the leading carriage seems to be an ex-NER non corridor! Most trains on the Edinburgh-Aberdeen route changed locomotives at Dundee to keep engines and men together – see page 41 – and this is an example of such an event. Note the steam driven generator for the electric lights behind the smoke deflector to the left of the smokebox and the BTH speed indicator driven off the trailing coupled axle. (Gavin L. Wilson/David Jenkinson Collection)

CHAPTER FOUR

Tailpiece

We conclude our review of the Thompson and Peppercorn pacifics with a brief view of their 'foreign' workings and a final look at what might almost be termed 'the first and the last'. Sadly in spite of their numbers, the last generation LNER 4-6-2s did not attract the attention of preservationists (official or otherwise) and at one time it seemed as though none at all would survive. In the case of the Thompson rebuilds, and even his A2/3s, this was perhaps understandable; they were not the brightest stars in the firmament! But the Peppercorn 4-6-2s merited a rather better fate than in fact they achieved.

The A1 type was a truly splendid design. Outperformed in power output terms only by the Stanier 'Duchesses' (and even then arguably at some cost in terms of economy and maintenance), the A1 could certainly lay sound claim to be not only the last of the purely 'express' steam engines to be built in quantity for service in Britain, but also one of the most outstanding – and it would be nice to have seen one saved. Fortunately, however, one of the substantially similar A2s has survived in the shape of *Blue Peter* and for this we must be grateful. Thus far it has seen little in the way of public exposure; but times may change!

Right: A Peppercorn A1 on the West Coast Main Line in early BR days: No. 60161 *North British* (12/49-10/63) – one wonders what the men of Polmadie with their Caledonian and LMS pedigree thought of a locomotive of *that* name and origin working over *their* route at that time – tops Beattock bank with the 11.15 am Birmingham-Glasgow express on 2nd May 1953. It would have come onto the train at Crewe, departing at 12.50 pm. The formation is eight coaches for Glasgow and four for Edinburgh. In June 1953 the two A1s remaining at Polmadie were returned to Haymarket. They made an interesting contrast to the Stanier Pacifics and represented an element that was missing from the 1948 Interchange Trials. They were lighter on coal and water and displayed lower maintenance costs, but perhaps they lacked the sheer brute strength of the 'Duchesses'. It was to be another ten years before the LNER breed was again to challenge the LMS product on its 'home' territory, so to speak, and opinions will always remain divided on this point! (J. Robertson/Colour Rail)

Left: Far from its home territory, Class A1 No. 60114 *W. P. Allen* (8/48-12/64) of Doncaster shed (36A) stands in Birmingham New Street platform 8 on a damp 8th August 1964. The train is 1N72 the 11.41 am Birmingham to Newcastle – a relief working to 1N28 the 9.20 am Cardiff to Newcastle. Rebuilding of the station had already begun and the southernmost bay of the Midland side roof had been demolished. The shed code of the engine seems irrelevant in context, and one can only presume that by this time, 'anything goes' was the underlying philosophy! (John Edgington)

On Sunday 16th July 1950, a special run was organised to celebrate the centenary of the opening of the Great Northern main line. The locomotive chosen was, appropriately, Class A1/1 No. 60113 *Great Northern* and a portrait of Edmund Beckett Denison, a founding father of the GNR, was displayed. The train ran between Kings Cross and York taking the old route via Boston, Lincoln, Retford, Knottingley and Burton Salmon on the outward journey. The first view shows No. 60113 at Lincoln on the down trip. The locomotive returned from York using the East Coast route and the second picture shows the locomotive at Kings Cross having completed the journey despite running into some crossing gates near Balne, between Selby and Doncaster.

Purists would doubtless argue that in rebuilt form, No. 60113 was not perhaps the most appropriate engine to have been chosen – name notwithstanding – but viewed with the hindsight of a further 40 years, maybe it was not bad decision. The arguments as to whether Edward Thompson was right or not to have rebuilt the celebrated pioneer Gresley 4-6-2 as he did, will probably continue for as long as folk have the energy to debate the issue; but what cannot be gainsayed is that emerging from the rebuilt *Great Northern,* came one of the most outstanding 4-6-2 designs to be built in this country – the Peppercorn A1. The real tragedy is that not a single example of this outstanding design has survived for posterity to examine – and no matter what one's emotional sentiments may be, this can only be regarded as a grave omission. (J. C. Flemmons/Millbrook House Collection)

Above: The last survivor working over Caledonian metals: No. 60532 *Blue Peter* (3/48-12/66) climbs Cumbernauld Glen with an eight coach Aberdeen-Glasgow (Buchanan Street) express on 23rd July 1966. The engine is allocated to Dundee (62B), shown LNER fashion on the bufferbeam, but the shed plate reads (61B) Aberdeen. The remaining A2s, like the Gresley A4s, had a last glimpse of express working over this route in the summer of 1966. The last two A4s were withdrawn in September of that year, so *Blue Peter* became the final LNER Pacific to be withdrawn from active service – on 31st December. Happily it was subsequently preserved as the only post-Gresley LNER 4-6-2 to be so distinguished. At the time of writing, the engine has only been seen (historically incorrectly) as LNER No. 532, but no matter; it still survives! (Maurice S. Burns)

BIBLIOGRAPHY

Locomotives of the LNER, RCTS, various parts 1963-1990.

Railway Clearing House Junction Diagrams, RCH 1928.

Gradient profiles of the British Main Line Railways, The Railway Publishing Co. 1936.

SLS Journal, various issues 1948-1963.

Railway Magazine, various issues 1948-1963.